VOLUME FIFTY SEVEN

ADVANCES IN
HEAT TRANSFER

VOLUME FIFTY SEVEN

Advances in
HEAT TRANSFER

Edited by

JOHN PATRICK ABRAHAM
School of Engineering,
University of St. Thomas,
St. Paul, MN, United States

JOHN M. GORMAN
Department of Mechanical Engineering,
University of Minnesota,
Minneapolis, MN, United States

WALLY J. MINKOWYCZ
University of Illinois,
Chicago, United States

Academic Press is an imprint of Elsevier
125 London Wall, London, EC2Y 5AS, United Kingdom
50 Hampshire Street, 5th Floor, Cambridge, MA 02139, United States
525 B Street, Suite 1650, San Diego, CA 92101, United States

First edition 2024

Copyright © 2024 Elsevier Inc. All rights are reserved, including those for text and data mining, AI training, and similar technologies.

Publisher's note: Elsevier takes a neutral position with respect to territorial disputes or jurisdictional claims in its published content, including in maps and institutional affiliations.

No part of this publication may be reproduced or transmitted in any form or by any means, electronic or mechanical, including photocopying, recording, or any information storage and retrieval system, without permission in writing from the publisher. Details on how to seek permission, further information about the Publisher's permissions policies and our arrangements with organizations such as the Copyright Clearance Center and the Copyright Licensing Agency, can be found at our website: www.elsevier.com/permissions.

This book and the individual contributions contained in it are protected under copyright by the Publisher (other than as may be noted herein).

Notices

Knowledge and best practice in this field are constantly changing. As new research and experience broaden our understanding, changes in research methods, professional practices, or medical treatment may become necessary.

Practitioners and researchers must always rely on their own experience and knowledge in evaluating and using any information, methods, compounds, or experiments described herein. In using such information or methods they should be mindful of their own safety and the safety of others, including parties for whom they have a professional responsibility.

To the fullest extent of the law, neither the Publisher nor the authors, contributors, or editors, assume any liability for any injury and/or damage to persons or property as a matter of products liability, negligence or otherwise, or from any use or operation of any methods, products, instructions, or ideas contained in the material herein.

ISBN: 978-0-443-29536-2
ISSN: 0065-2717

For information on all Academic Press publications
visit our website at https://www.elsevier.com/books-and-journals

Publisher: Zoe Kruze
Acquisitions Editor: Leticia M. Lima
Editorial Project Manager: Devwart Chauhan
Production Project Manager: Abdulla Sait
Cover Designer: Christian Bilbow

Typeset by MPS Limited, India

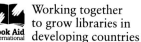

Contents

Contributors	*vii*
Preface	*ix*

1. Review of the cavity effect: Modeling and impact of cavity shape on apparent radiative surface properties — **1**

Ehsan Mofidipour, Matthew R. Jones, and Brian D. Iverson

1. Introduction	3
2. Objective and overview	5
3. General model for the spectral, hemispherical apparent emissivity of a cavity	6
4. Verification of the general model for apparent radiative surface properties	15
5. Apparent emissivity for isothermal, diffuse, gray, two-surface enclosures	20
6. Advances in cavity modeling	29
7. Comparison of hemispherical apparent emissivity	42
8. Cavity effect applications	54
9. Conclusions	59
Acknowledgments	61
References	61

2. Hyperthermia applications in cardiovascular and cancer therapy treatments — **71**

Sanaz Imanlou and Kambiz Vafai

1. Introduction	72
2. Hyperthermia in cardiovascular diseases	72
3. Using hyperthermia in cancer treatment applications	79
4. Conclusions	91
References	92

3. Enhancement of heat transfer with nanofluids and its applications in heat exchangers — **101**

Wajahat Ahmed Khan, Kaleemullah Shaikh, Rab Nawaz, Salim Newaz Kazi, and Mohd Nashrul Mohd Zubir

1. Introduction	102
2. Application of nanofluids in solar collectors	105

3.	Application of nanofluids for fouling retardation and heat transfer enhancement	111
4.	Applications of nanofluids in annular flow heat exchangers	118
5.	Conclusion	123
	Acknowledgements	124
	References	124

4. Programmable micro- and nanoengineered liquid metals in thermal engineering applications
129

Rahul Agarwal, Saleh S. Baakeem, and A.A. Mohamad

1.	Introduction	130
2.	Morphological forms	132
3.	Processing techniques	134
4.	Actuation principles and associated characteristics of liquid metals	137
5.	Applications of liquid metals in thermal engineering	156
6.	Our perspective and future directions	173
	References	174

5. Thermal transport in engineered cellular materials: A contemporary perspective
183

Prashant Singh and Roop L. Mahajan

1.	Introduction	184
2.	Stochastic foams	185
3.	Pore-scale flow and thermal transport in stochastic cellular materials	195
4.	Pore-scale flow and thermal transport in architectured cellular materials: strut-based	205
5.	Concluding remarks	227
	References	229

6. Heat transfer analysis of partially ionized hybrid nanofluids flow comprising magnetic/non-magnetic nanoparticles in an annular region of two homocentric inclined cylinders
237

Muhammad Ramzan and Nazia Shahmir

1.	Introduction	238
2.	Mathematical formulation	241
3.	Conclusions	252
	References	252

Contributors

Rahul Agarwal
Department of Mechanical and Manufacturing Engineering, University of Calgary, Calgary, AB, Canada

Saleh S. Baakeem
Department of Chemical and Petroleum Engineering, University of Calgary, Calgary, AB, Canada

Sanaz Imanlou
Mechanical Engineering Department, University of California, Riverside, CA, United States

Brian D. Iverson
Brigham Young University, Provo, UT, United States

Matthew R. Jones
Brigham Young University, Provo, UT, United States

Salim Newaz Kazi
Department of Mechanical Engineering, Faculty of Engineering, Universiti Malaya, Kuala Lumpur, Malaysia

Wajahat Ahmed Khan
Nanotechnology & Catalysis Research Centre (NANOCAT), Institute for Advanced Studies (IAS), Universiti Malaya, Kuala Lumpur, Malaysia

Roop L. Mahajan
Department of Mechanical Engineering, Virginia Tech, Blacksburg, VA, United States

Ehsan Mofidipour
Brigham Young University, Provo, UT, United States

A.A. Mohamad
Department of Mechanical and Manufacturing Engineering, University of Calgary, Calgary, AB, Canada

Mohd Nashrul Mohd Zubir
Department of Mechanical Engineering, Faculty of Engineering, Universiti Malaya, Kuala Lumpur, Malaysia

Rab Nawaz
Department of Mechanical Engineering, Faculty of Engineering, Universiti Malaya, Kuala Lumpur, Malaysia

Muhammad Ramzan
Department of Computer Science, Bahria University, Islamabad, Pakistan

Nazia Shahmir
Department of Computer Science, Bahria University, Islamabad, Pakistan

Kaleemullah Shaikh
Department of Mechanical Engineering, Faculty of Engineering, Universiti Malaya, Kuala Lumpur, Malaysia

Prashant Singh
Department of Mechanical, Aerospace & Biomedical Engineering, University of Tennessee, Knoxville, TN, United States

Kambiz Vafai
Mechanical Engineering Department, University of California, Riverside, CA, United States

Preface

Volume 57 of *Advances in Heat Transfer* continues a long tradition of presenting significant contributions from leaders in the thermal science community. This series publishes work that is more detailed and expansive than typical journal papers. It allows authors to explore topics more richly and to provide context for their work. In this issue, topics include radiative heat transfer in cavities, biomedical applications of hyperthermia, improvements to heat transfer through the use of nanofluids, the use of liquid metals in heat transfer, thermal transport in cellular materials, and advancements in nanofluids including ionized hybrid nanofluids with and without magnetic forces.

In Chapter 1, radiative heat transfer is the focus, and the influence of cavity shape is investigated. First, a review of radiation in cavities is provided. Then, a generalized model of generic cavities is developed using the apparent emissivity concept. Next, the authors discuss verification methods for surface properties and apply their apparent emissivity to isothermal, diffuse, gray, and two-surface cavities. Numerous different cavity shapes are included, and new techniques for cavity modeling are articulated. This chapter concludes with a discussion of applications that provide ample motivation for the chapter.

Chapter 2 deals with biomedical applications of elevated tissue temperature (hyperthermia). While elevated temperatures affect tissue temperature, the sensitivity of tissue to temperatures and the impact of elevated temperatures on therapy is not fully known. The authors focus their attention on cardiovascular and cancer applications. Concerning cardiovascular applications, hyperthermia influences mass transport across tissue layers and subsequently on hypertension and LDL concentrations. For cancer therapies, the chapter utilizes various nanoparticles that are used to target thermal therapy (gold, magnetic, iron, and carbon nanoparticles). The authors further explore how the presence of blood vessels can alter tissue temperatures. Blood vessels can potentially limit hyperthermic treatments if not accounted for. The chapter ends with a discussion of future research areas on this topic.

Chapter 3 explores nanofluid use in heat exchangers. It is well known that nanoparticles, when suspended in a base fluid, can significantly alter the thermal conductivity and viscosity. These changes affect both heat transfer and flow efficiency. Typically, nanofluids increase heat transfer but reduce flow efficiency, so a cost-benefit analysis is required. The authors apply nanofluids to numerous heat exchangers, including direct absorption, flat plate, and evacuated tube solar collectors. The authors also investigate fouling, which can occur at the bounding walls and impede further flow. Lastly, the authors incorporate nanofluids into annular heat exchangers.

Metallic fluids are the focus of Chapter 4. Multiple morphologies are considered (bulk droplets, particles, and liquid metal marbles). The authors discuss processing techniques, including drop-on-demand, molding, microfluidics, sonic, shearing, and liquid metal marbles. The authors also showcase different actuation methods (mechanical, electrical, magnetic, optical, acoustic, thermal, chemical, etc.). Last, the authors go into great detail, showcasing different applications of liquid metals in the thermal sciences (phototherapeutic, thermal switches, actuators, energy harvesting, microfluidics, thermal interfaces, composite materials, and phase change materials). The chapter concludes with a discussion of future research areas in this discipline.

In Chapter 5, we find a contribution to thermal transport in cellular materials. These cellular materials include metal foams known for their excellent thermal transport ability. Foams have very high area-to-volume ratios, and the porous materials promote flow mixing. The authors artfully describe different types of porous substructures and the relationship between thermal transport, permeability, and porosity. The authors create pore-scale computational models with incredible details of the sub-pore flow patterns.

The last chapter is another contribution to nanofluids. Specifically, it deals with partially ionized hybrid nanofluids with and without magnetic effects flowing in an annular space. The authors can reduce the governing differential equations to a dimensionless form that allows calculation. Hall effect and ion slip are considered, and the nanoparticles are made from silver, copper, cobalt ferrite, and manganese ferrite. The solutions provided by the authors allow users to tailor fluidic parameters to meet the needs of case-specific applications.

These six chapters are contributed by experts from across the globe. *Advances in Heat Transfer* strives to continue to provide in-depth reviews and emerging science in various thermal science topics. We will continue this tradition in the coming years.

Editors
JOHN PATRICK ABRAHAM
JOHN M. GORMAN
WALLY J. MINKOWYCZ

CHAPTER ONE

Review of the cavity effect: Modeling and impact of cavity shape on apparent radiative surface properties

Ehsan Mofidipour, Matthew R. Jones, and Brian D. Iverson[*]
Brigham Young University, Provo, UT, United States
*Corresponding author. e-mail address: bdiverson@byu.edu

Contents

1. Introduction	3
2. Objective and overview	5
3. General model for the spectral, hemispherical apparent emissivity of a cavity	6
3.1 Spectral, hemispherical apparent emissivity of a cavity	7
3.2 Apparent emissivity of a cavity with isothermal, diffuse, and gray walls	9
3.3 Equivalence of apparent emissivity and absorptivity	11
3.4 Hemispherical, integrated, and local apparent emissivity	12
3.5 Methods for determining apparent radiative surface properties	13
4. Verification of the general model for apparent radiative surface properties	15
5. Apparent emissivity for isothermal, diffuse, gray, two-surface enclosures	20
5.1 Cylindrical cavity	21
5.2 Conical cavity	22
5.3 Spherical cavity	23
5.4 Cylindro-conical cavity	23
5.5 Cylindro-inner cone cavity	24
5.6 Double-cone cavity	24
5.7 Geometric comparison	25
6. Advances in cavity modeling	29
6.1 Cylindrical cavity	30
6.2 Conical cavity	36
6.3 Spherical cavity	38
6.4 Cylindro-conical cavity	39
6.5 Cylindro-inner cone cavity	41
6.6 Double-cone cavity	42
7. Comparison of hemispherical apparent emissivity	42
8. Cavity effect applications	54
9. Conclusions	59
Acknowledgments	61
References	61

Advances in Heat Transfer, Volume 57
ISSN 0065-2717, https://doi.org/10.1016/bs.aiht.2023.12.002
Copyright © 2024 Elsevier Inc. All rights are reserved, including those for text and data

Abstract

Radiative surface properties play a critical role in the analysis, design, and optimization of thermal systems. Geometry has a strong influence on the emission and absorption characteristics of a surface. Modification of radiative surface properties may be achieved by using engineered surfaces or cavities by capitalizing on the cavity effect. A model for the apparent emissivity of an arbitrarily shaped cavity with spectrally and directionally dependent radiative surface properties is presented in this work. This model is verified through comparison with published models of the apparent radiative properties of cylindrical cavities. The merit of this model is demonstrated by obtaining closed-form expressions for the apparent emissivity of ideal, two-surface cavities for six representative geometries (cylindrical, conical, spherical, cylindro-conical, cylindro-inner cone, and double-cone). A comprehensive review of prior research related to the apparent emissivity of these shapes is presented. Results obtained using the simple, closed-form expressions agree well with results obtained using more complex methods when the intrinsic surface emissivity is large. Finally, this work elaborates on applications of radial and angular cavity shapes and describes the potential benefits of using geometric manipulation for dynamic control of radiative surface properties in energy and thermal management systems.

Abbreviations

A	Surface area [m^2].
d	Double-cone cavity diameter [m].
D	Diameter [m].
E	Emissive power [W m^{-2}].
f	Fraction of irradiation reflection [−].
G	Irradiation [W m^{-2}].
h	Conical height [m].
I	Intensity [W m^{-2} sr^{-1}].
J	Radiosity [W m^{-2}].
L	Length [m].
q	Heat transfer rate [W].
q''	Heat flux [W m^{-2}].
$\vec{n}$	Surface normal vector [−].
M	Number of surface elements [−].
R	Reference surface area [m^2].
S	Distance between two surface elements [m].
T	Temperature [K].
θ	Angle between the emission and the normal vector [°].
λ	Wavelength [μm].
ε	Intrinsic emissivity [−].
α	Intrinsic absorptivity [−].
ρ	Reflectivity [−].
ρ''	Bidirectional reflection function [sr^{-1}].
γ	Collimated irradiation angle [°].
Ω	Solid angle [sr].
σ	Stefan-Boltzmann constant [W m^{-2} K^{-4}].

Subscripts

$[]_a$	Indicates that [] is an apparent quantity.
$[]_b$	Indicates that [] is an ideal blackbody quantity.
$[]_c$	Indicates that [] is associated with the conical section.
$[]_e$	Indicates that [] is an emitted quantity.
$[]_{eff}$	Indicates that [] is an effective quantity.
$[]_i$	Indicates that [] is an incident quantity.
$[]_{inc}$	Indicates that [] is associated with the inclined bottom cavity.
$[]_m$	Indicates that [] is associated with surface m.
$[]_n$	Indicates that [] is reflected n times.
$[]_s$	Indicates that [] is associated with the spherical cavity.
$[]_{surr}$	Indicates that [] is a surrounding quantity.
$[]_\theta$	Indicates that [] is directionally dependent.
$[]_\lambda$	Indicates that [] is spectrally dependent.
$[]_0$	Indicates that [] is associated with the cavity opening.
$[]_1$	Indicates that [] is associated with the cavity wall.

1. Introduction

Emission and absorption of thermal radiation are essential factors in the design of thermal management systems. The net radiative heat exchange at a surface is determined by its radiative surface properties, and engineering a surface such that it has specified radiative surface properties proves advantageous in many applications [1–4]. Radiative surface properties depend strongly on surface geometry, so desired radiative surface properties may be achieved by patterning a surface with grooves or indentations. This phenomenon is referred to as the cavity effect.

Consider the cavity shown in Fig. 1A with a small opening through which radiation can enter or leave. An imaginary planar surface covering the opening of a cavity is referred to as the aperture, and the effective radiative properties of the aperture are the 'apparent' radiative surface properties of the cavity. The cavity effect and the nomenclature associated with the aperture are shown in Fig. 1. Rays representing radiative energy may leave the cavity after multiple reflections [5] and the apparent radiative surface properties may approach those of a blackbody.

Radiation entering the cavity may be reflected multiple times. Since some absorption occurs during each interaction with the cavity surface, the apparent absorptivity of the aperture is larger than the intrinsic absorptivity of the cavity surfaces. Likewise, emission from interior cavity walls may be reflected multiple times within the cavity, resulting in diffuse emission and

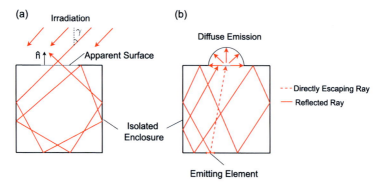

Fig. 1 Multiple interactions between the irradiation and the cavity walls increase the apparent absorptivity of the surface. (A) Illustration of a cavity exposed to irradiation from an external source. (B) Illustration of emission from a cavity. Radiation reflected and re-emitted by the cavity surfaces increases the emissive power passing through the aperture, which increases the apparent emissivity of the opening.

an apparent emissivity greater than the intrinsic emissivity of the cavity surfaces (Fig. 1B). In this manner, the apparent emissivity of the aperture may be tuned by varying the geometry of the cavity. As such, the apparent emissivity may approach that of a blackbody or some value greater than the intrinsic emissivity of the cavity surfaces [6].

Since the intrinsic radiative properties of a surface are usually fixed, they cannot be adjusted in response to changes in the environment or operating conditions. Various technologies have been suggested to dynamically alter radiative properties, including surface coatings, electrowetting, and electrostatic actuation [7–12]. Although effective, these approaches may adjust to a changing environment or operating condition slowly or may be limited in their range. An approach that may have a large dynamic range and be able to respond rapidly to variations in the environment or operating conditions is to modify the topology of a surface by geometric manipulation [9,13–15]. Altering the shape of the surface thereby offers the means to control radiative heat exchange in a dynamic fashion. This control helps maintain the temperature of the working components at the desired level. Systems or equipment that require thermal management are ubiquitous. Two notable examples occur in spacecraft thermal management [16] (Fig. 2A), where excessive heat gain and loss should be avoided, and in particle flow solar reactors (Fig. 2B), where a cylindrical cavity is employed to effectively collect solar energy for the co-production of hydrogen and carbon [17]. Additional applications involving the cavity effect are described in Section 8.

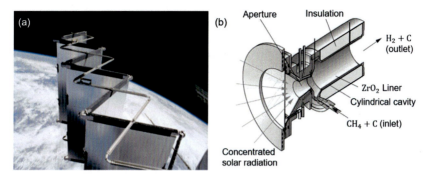

Fig. 2 Examples of cavity geometric shapes that can appear in (A) thermal management of spacecraft radiators [16] and in (B) collection of solar energy with a cylindrical cavity solar reactor for the co-production of hydrogen and carbon. *Adapted from A. Steinfeld, Solar thermochemical production of hydrogen - a review, Sol. Energy. 78 (2005) 603–615. https://doi.org/10.1016/j.solener.2003.12.012.*

Since the closest approximation of a perfect blackbody is an isothermal cavity with a small opening, there have been significant efforts to create blackbody absorption or emission using different cavity geometric shapes [18,19]. Such cavities have been used for calibration equipment or creating radiation sources that are well characterized. With geometric manipulation, one can effectively regulate the radiative behavior of components in the transient thermal management of objects operating in different applications. This study explores the effect of geometry on cavity design and summarizes corresponding works published by other researchers.

2. Objective and overview

Review papers explaining different methods that have been employed to obtain the apparent radiative surface properties of cavities exist in the literature [20,21]. These papers present excellent discussions regarding the significance of cavity surface properties, the nature of emission and reflection from the cavity, spectral range, and working environment in modeling thermal radiation. Three common cavity shapes (e.g. cylinder, cone, and sphere) have primarily been considered in these works. The present work is complementary to these review papers since it focuses on the impact of altering the geometric shape of a cavity over a broader range of surface conditions or properties.

This work provides a general formulation for the apparent emissivity of an arbitrarily shaped cavity. The general equation can be employed for non-isothermal cavities with varying intrinsic spectral, directional radiative surface properties, including diffuse or specular reflectivity. We verify this general approach by determining the total, hemispherical apparent emissivity of a cylindrical cavity through comparison with results available in the literature.

Furthermore, this general formulation is used to obtain closed-form expressions of the total, hemispherical apparent emissivity of two-surface enclosures with isothermal, opaque, diffuse, and gray walls for six isothermal cavity shapes, including a cylinder, cone, sphere, cylindro-cone, cylindro-inner cone, and double-cone. Analytical expressions for the hemispherical apparent emissivity of the cylindro-cone, cylindro-inner cone, and double-cone under these conditions have not previously been reported in the literature. In cases where a model for hemispherical emissivity has been published, validation was performed by comparison with available results. Expressions for the total, hemispherical apparent emissivity for the six cavity shapes demonstrate how the general formulation may be applied to any cavity geometry.

Finally, this study summarizes existing literature regarding the measurement or computation of the apparent emissivity and explores the significance of different geometries. As a part of this literature review, we identify analyses and assumptions employed to characterize apparent radiative properties. Experimental, analytical, and computational studies are organized by geometry, providing a comprehensive collection of related works for a given cavity shape. The general model presented in this paper serves as a tool for determining the apparent radiative surface properties of an arbitrarily shaped cavity with varying intrinsic radiative surface properties and wall temperature profiles. Alternatively, the general model may be used to design cavities that result in desirable apparent radiative surface properties for a given application.

3. General model for the spectral, hemispherical apparent emissivity of a cavity

In this section, a general formulation for the apparent emissivity of an arbitrarily shaped cavity is developed. As shown in Section 5, this general formulation may be evaluated analytically in some cases. The apparent emissivity is determined by modeling the radiation exchange between surfaces comprising the cavity. In some published literature, radiative exchange between surfaces is modeled using the standard image method [22], often limiting their analysis to nominally flat surfaces. For a complex geometry, the standard image method is not applicable due to the shadowing effect [23]. Shadowing occurs when some surfaces are blocked from the direct line of sight from other surfaces. The standard image method is also not able to account for multiple reflections inside the cavity.

3.1 Spectral, hemispherical apparent emissivity of a cavity

We first develop an expression for the apparent emissivity of an opaque cavity with spectrally and directionally dependent intrinsic radiative surface properties. As illustrated in Fig. 3, the interior surface of an arbitrarily shaped cavity is comprised of M differential surfaces. The medium filling the cavity is non-participating.

Consistent with the definition of the spectral, hemispherical emissivity [24,25], the spectral, hemispherical apparent emissivity of the cavity is defined as the ratio of the cavity's spectral emissive power to the spectral emissive power of a blackbody at the effective temperature of the cavity, T_{eff}.

$$\varepsilon_{\lambda,a} = \frac{E_\lambda}{E_{\lambda,b}(T_{eff})} \quad (1)$$

The spectral emissive power leaving the cavity is equal to the spectral irradiation incident on an imaginary surface, ΔA_0.

$$E_\lambda = G_{\lambda,0} = \int_{2\pi} I_{\lambda,i,0} \cos\theta \, d\Omega \quad (2)$$

Neglecting any participating medium, the spectral intensity incident on the opening (ΔA_0) from the direction specified by the solid angle subtended

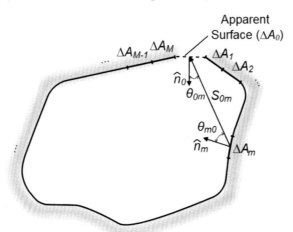

Fig. 3 Illustration of an arbitrary cavity comprised of M differential surfaces.

by a differential element of the cavity surface (ΔA_m) is equal to the spectral intensity emitted and reflected by ΔA_m in the direction of the opening. Therefore, the spectral irradiation incident on the opening is as follows.

$$G_{\lambda,0} = \sum_{m=1}^{M} \int_{\Delta\Omega_{0\to m}} I_{\lambda,e+r,m} \cos\theta_{0m}\, d\Omega \tag{3}$$

The spectral intensity emitted and reflected from surface m toward the cavity opening is

$$I_{\lambda,e+r,m} = \varepsilon_{\lambda,\theta_{m0}} I_{\lambda,b}(T_m) + \sum_{m=1}^{M} \int_{2\pi} \rho_{\lambda}''(\Omega \to \Delta\Omega_{m\to0}) I_{\lambda,i,m} \cos\theta\, d\Omega. \tag{4}$$

The first term on the right-hand side of Eq. (4) represents the spectral intensity emitted by surface m within the solid angle subtended by the cavity opening. The spectral, directional emissivity of surface m, $\varepsilon_{\lambda,\theta_{m0}}$, is an intrinsic property. The second term on the right-hand side of Eq. (4) represents the spectral intensity reflected from surface m in the direction of the opening due to incident radiation from all directions. The spectral, bidirectional reflectivity of surface m, $\rho_{\lambda}''(\Omega \to \Delta\Omega_{m\to0})$, is also an intrinsic radiative surface property that represents the fraction of spectral intensity incident on surface m from the Ω direction that is reflected in the direction of the opening.

The spectral emissive power of a blackbody at the effective temperature of the cavity (denominator of Eq. (1)) is found by approximating each differential surface comprising the interior of the cavity as a black surface. Following the same logic used to develop Eq. (3), the spectral emissive power of a blackbody cavity is

$$E_{\lambda,b}(T_{eff}) = \sum_{m=1}^{M} \int_{d\Omega_{0\to m}} I_{\lambda,b}(T_m) \cos\theta_{0m}\, d\Omega \tag{5}$$

Substituting Eqs. (3–5) into Eq. (1) gives the following general expression for the spectral, hemispherical apparent emissivity of an arbitrarily shaped, non-isothermal cavity with spectrally and directionally dependent radiative surface properties.

$$\varepsilon_{\lambda,a} = \frac{\sum_{m=1}^{M}\int_{\Delta\Omega_{0\to m}}\left[\varepsilon_{\lambda,\theta_{m0}}I_{\lambda,b}(T_m) + \sum_{m=1}^{M}\int_{2\pi}\rho_{\lambda}''(\Omega \to \Delta\Omega_{m\to0})I_{\lambda,i,m}\cos\theta\, d\Omega\right]\cos\theta_{0m}\, d\Omega}{\sum_{m=1}^{M}\int_{d\Omega_{0\to m}}I_{\lambda,b}(T_m)\cos\theta_{0m}\, d\Omega} \tag{6}$$

3.2 Apparent emissivity of a cavity with isothermal, diffuse, and gray walls

Eq. (6) is a general expression for the spectral, hemispherical apparent emissivity of a non-isothermal cavity whose inner surface is comprised of opaque surfaces with directionally and spectrally varying intrinsic radiative surface properties. Further analysis of the spectral, hemispherical apparent emissivity based on this general expression requires specification of the temperature and the spectral bidirectional reflectivity of each differential surface area that comprises the interior of the cavity. Since the focus of this work is primarily on the effect of a cavity's geometry on the spectral, hemispherical apparent emissivity, we now focus on isothermal cavities with diffuse intrinsic radiative surface properties. We also assume that the intrinsic radiative surface properties of each of the M differential surfaces that comprise the interior surface of the cavity are the same.

Since the differential surfaces reflect diffusely [24],

$$\rho_\lambda'' (\Omega \rightarrow \Delta\Omega_{m\to 0}) = \frac{\rho_\lambda}{\pi}. \tag{7}$$

Substituting Eq. (7) into Eq. (4) and the resulting equation into Eq. (3) gives the spectral irradiation incident on the opening.

$$G_{\lambda,0} = \sum_{m=1}^{M} \left[\int_{\Delta\Omega_{0\to m}} \varepsilon_\lambda I_{\lambda,b}(T_m) + \frac{1}{\pi} \int_{2\pi} \rho_\lambda I_{\lambda,i,m} \cos\theta d\Omega \right] \cos\theta_{0m} d\Omega \tag{8}$$

Since each differential surface is diffuse, the spectral intensity incident on any differential surface m will also be approximately diffuse. Therefore, Eq. (8) simplifies as follows.

$$G_{\lambda,0} = \sum_{m=1}^{M} \int_{\Delta\Omega_{0\to m}} \left(\varepsilon_\lambda I_{\lambda,b}(T_m) + \left(\frac{1-\varepsilon_\lambda}{\pi} \right) I_{\lambda,i,m} \right) \cos\theta_{0m} d\Omega \tag{9}$$

Multiplying the right side of Eq. (9) by $\frac{\pi\Delta A_0}{\pi\Delta A_0}$ and using the definition of the solid angle subtended by the differential element, m, when viewed from the cavity opening gives

$$G_{\lambda,0} = \sum_{m=1}^{M} \pi \left(\varepsilon_\lambda I_{\lambda,b}(T_m) + \left(\frac{1-\varepsilon_\lambda}{\pi} \right) I_{\lambda,i,m} \right)$$

$$\frac{1}{\Delta A_0} \int_{\Delta A_0} \int_{\Delta A_m} \frac{\cos\theta_{0m} \cos\theta_{m0}}{\pi S_{0m}^2} dA_m dA_0. \tag{10}$$

Using definitions of the spectral radiosity and diffuse view factor, Eq. (10) becomes

$$G_{\lambda,0} = \sum_{m=1}^{M} F_{0m} J_{\lambda m}. \tag{11}$$

Similarly, Eq. (5) simplifies as follows.

$$\begin{aligned}
E_{\lambda,b}(T_{\text{eff}}) &= \sum_{m=1}^{M} \pi I_{\lambda,b}(T_m) \frac{1}{\Delta A_0} \int_{\Delta A_0} \int_{\Delta A_m} \frac{\cos\theta_{0m} \cos\theta_{m0}}{\pi S_{0m}^2} dA_m \, dA_0 \\
&= \sum_{m=1}^{M} F_{0m} E_{\lambda,b}(T_m) \tag{12}
\end{aligned}$$

Substituting Eqs. (11) and (12) into Eq. (1) results in an expression for the spectral, hemispherical apparent emissivity of a cavity with diffusely reflecting and emitting walls.

$$\varepsilon_{\lambda,a} = \frac{\sum_{m=1}^{M} F_{0m} J_{\lambda m}}{\sum_{m=1}^{M} F_{0m} E_{\lambda,b}(T_m)} \tag{13}$$

It should be noted that Eq. (13) is strictly valid only when the cavity walls are diffusely emitting and reflecting. However, the radiation field in a cavity is often diffuse due to multiple reflections, so the spectral, hemispherical apparent emissivity may be accurately modeled using Eq. (13) even when the intrinsic radiative surface properties of the cavity are directionally dependent. The effect of directionally dependent properties on the apparent radiative properties of a cavity will be further investigated in Section 7.

If the cavity walls are isothermal, T_m is the same for each differential surface, and Eq. (13) simplifies to

$$\varepsilon_{\lambda,a} = \frac{\sum_{m=1}^{M} F_{0m} J_{\lambda m}}{E_{\lambda,b}(T_m) \sum_{m=1}^{M} F_{0m}} = \frac{\sum_{m=1}^{M} F_{0m} J_{\lambda m}}{E_{\lambda,b}(T_m)} \tag{14}$$

where the summation rule [25] was used to simplify Eq. (14). Here, it is useful to note that if the cavity is isothermal, the temperature that characterizes emission from the cavity (i.e. the effective temperature, T_{eff}) is equal to the temperature of the differential element surfaces, T_m.

$$E_b(T_{\text{eff}}) = \sigma T_{\text{eff}}^4 = \int_0^\infty \int_{2\pi} I_{\lambda,i,0} \cos\theta \, d\Omega \, d\lambda \tag{15}$$

If all surfaces comprising the cavity are black, then

$$I_{\lambda,i,0} = I_{\lambda,b}(T_m) \tag{16}$$

for all incident directions. Under these conditions, substituting Eq. (16) into Eq. (15) gives

$$\sigma T_{eff}^4 = \int_0^\infty \sum_{m=1}^M \int_{\Delta\Omega_{0\to m}} I_{\lambda,b}(T_m) \cos\theta_{0m} \, d\Omega \, d\lambda \tag{17}$$

Following the same procedure as employed when simplifying Eqs. (3) and (5), Eq. (17) becomes

$$\sigma T_{eff}^4 = \int_0^\infty \sum_{m=1}^M \pi I_{\lambda,b}(T_m) \left[\frac{1}{\Delta A_0} \int_{\Delta A_0} \int_{\Delta A_m} \frac{\cos\theta_{0m}\cos\theta_{m0}}{\pi S_{0m}^2} dA_m \, dA_0 \right] d\lambda$$

$$= \int_0^\infty E_{\lambda,b}(T_m) \sum_{m=1}^M F_{0m} \, d\lambda = \sigma T_m^4. \tag{18}$$

Therefore, the effective temperature is equal to the surface temperature of an isothermal cavity $(T_{eff} = T_m)$, as expected. The isothermal approximation may be unrealistic in some applications, but it facilitates comparison of the radiative behavior of different cavity shapes, which is a focus of this paper. The significance of assuming a non-isothermal cavity is discussed in Section 6. It should be noted that the following expression for $T_{eff} = \left[\frac{\sum_{m=1}^M F_{0m} E_b(T_m)}{\sigma} \right]^{\frac{1}{4}}$ can be employed to obtain the effective temperature in cases where the cavity is non-isothermal.

Integrating the expression for the spectral, hemispherical apparent emissivity (Eq. (14)) over the entire spectrum results in an expression for the total, hemispherical apparent emissivity.

$$\varepsilon_a = \frac{\int_0^\infty \sum_{m=1}^M F_{0m} J_{\lambda m} \, d\lambda}{\int_0^\infty E_{\lambda,b}(T_m) \, d\lambda} = \frac{\sum_{m=1}^M F_{0m} J_m}{\sigma T_m^4} \tag{19}$$

Eq. (19) is an expression for the total, hemispherical apparent emissivity (also called the effective emissivity in some papers [26,27]) of an isothermal cavity whose inner surface emits and reflects diffusely and is opaque. Using this relationship requires the determination of (1) the view factors between each interior cavity surface and the cavity opening and (2) the radiosity of each elemental surface. Along with providing this general formulation for the apparent emissivity for any arbitrary cavity, we apply this approach in Section 5 to six representative cavity shapes, which demonstrates the merit of the proposed formulation.

3.3 Equivalence of apparent emissivity and absorptivity

According to Kirchhoff's law, the spectral directional emissivity and absorptivity of a surface are equivalent $(\varepsilon_{\lambda,\theta} = \alpha_{\lambda,\theta})$. This condition is always applicable as they are inherent radiative surface properties and independent of the nature of the emission and reflection [25,28]. It may be shown that if surfaces are diffuse or the irradiation is diffuse, the spectral hemispherical emissivity is also equal to the spectral hemispherical absorptivity ($\varepsilon_\lambda = \alpha_\lambda$). If the surface is opaque and gray or the incoming irradiation is from a blackbody at the same temperature as the surfaces comprising the cavity, hemispherical intrinsic absorptivity and hemispherical intrinsic emissivity are also equivalent and they are both equal to one minus the hemispherical reflectivity ($\varepsilon = \alpha = 1 - \rho$).

Similar arguments may be used to show that the apparent absorptivity of a cavity aperture is equal to its apparent emissivity if the radiation within the cavity is diffuse. Therefore, the cavity walls may be specularly reflecting as long as the shape of the cavity results in a diffuse radiation field within the cavity [28]. While this work outlines the method to determine the hemispherical apparent emissivity of a cavity comprised of diffusely emitting surfaces, the outcome can also be used to obtain the apparent absorptivity of a cavity.

3.4 Hemispherical, integrated, and local apparent emissivity

Hemispherical, local, and integrated apparent emissivity are terms that are commonly used by researchers when cavity analyses are performed. The hemispherical apparent emissivity is defined as the ratio of total emission from a cavity opening to that of a blackbody cavity at an effective temperature for the cavity (Fig. 4A). The hemispherical apparent emissivity is a function of different geometric and thermal characteristics such as shape, surface roughness and temperature of surfaces comprising the cavity, which makes an accurate measurement of this quantity hard to perform. Depending on the need, the local quantity of the apparent emissivity may be favorable.

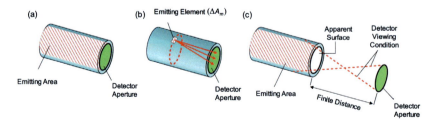

Fig. 4 The placement and viewing conditions of a detector to compute/measure the (A) hemispherical, (B) local, and (C) integrated apparent emissivity as discussed in the literature.

The local apparent emissivity is the ratio of the radiosity of a surface element ΔA_m at a specific location inside the cavity and temperature T_m to the emissive power of a blackbody cavity at an effective temperature (T_{eff}). In other words, in the local apparent emissivity calculation/measurement, incident radiation from a surface element specified on the interior of the cavity at the temperature T_m that is received by the detector is compared to that of a blackbody at an effective temperature (T_{eff}). The local apparent emissivity is used to identify the degree of uniformity of apparent emissivity at different locations and for different cavity shapes (Fig. 4B) [29]. Due to the non-uniformity of some cavities' fabrication, some detectors may receive irradiation from different parts of cavities that are at a lower temperature than the average cavity temperature [30]. In such scenarios, the radial distribution of the local apparent emissivity can play a vital role in enhancing the accuracy of radiative property calculations.

When the local apparent emissivity is integrated over the part of the cavity wall that is visible to a detector, which is placed at some finite distance from the cavity opening, the integrated emissivity is obtained (Fig. 4C) [31]. It is important to acknowledge that if the detector is placed on top of the cavity opening and its normal coincides with the cavity opening's normal, the hemispherical apparent emissivity is obtained. Fig. 4 illustrates the placement and viewing conditions associated with the detector and the emitting area in the measurement/computation of these forms of apparent emissivity. The 'normal' apparent emissivity (not shown in Fig. 4) is also a limiting case of the integrated emissivity, where the detector is placed at a sufficiently long distance from the cavity opening so as to determine the emission in the normal direction to the cavity opening. In the current work, we only formulate the hemispherical apparent emissivity, though we review in Section 6 the literature for different cavity shapes and include references to these different types of emissivity.

3.5 Methods for determining apparent radiative surface properties

Apparent radiative surface properties of various cavity shapes have been quantified using numerical/computational, experimental, and analytical methods. Numerical methods can be practical for analyzing complex cavity shapes since exact solutions of the governing integral equations for the apparent emissivity are difficult to obtain.

Monte Carlo Ray Tracing (MCRT) is one numerical approach that has been widely implemented for the quantification of radiative heat exchange between surfaces. MCRT is commonly formulated with one of two general algorithms [28,32,33]. The first approach uses reflectivity as the probability for reflection, with non-reflected rays terminating. In this approach, the algorithm estimates the probability of a random ray escaping from the cavity opening after a certain number of reflections within the cavity (emission scheme). The second approach tracks the reduction of the radiative energy of each ray, essentially decreasing the ray energy with each absorption/reflection event (absorption scheme). In the absorption scheme, the path of a discrete bundle of rays is followed until the energy assigned to each ray is negligibly small (i.e., fully absorbed by the cavity interior surfaces) or exits the cavity. The net heat exchange is then the difference between the number of rays absorbed and the number of bundles emitted at a given surface [34,35]. MCRT is a promising method to characterize the apparent radiative surface properties of cavities and can be applied to any arbitrarily shaped cavity as it has the flexibility of defining radiation exchange conditions as well as accommodating various interior surface properties. Methods for accelerating the probabilistic nature of MCRT analysis are also available in the literature [28].

Another approach to determining apparent radiative properties of cavities is through conducting experiments using an optical setup. The Fourier Transform Infrared Spectrometer (FTIR) and Integrating Sphere Reflectometer (ISR) shown in Fig. 5A and B are commonly employed instruments in radiative property measurements. The FTIR is a device that measures the spectral intensity of radiation incident on a detector installed within the FTIR. To determine apparent radiative properties, one may use an infrared light source or external optical fiber to irradiate samples in a sample compartment and obtain emission signals from the sample. Spectral emission from a sample or cavity is then compared with that of a blackbody reference using Planck's distribution to obtain apparent radiative properties [36,37]. When an FTIR spectrometer is used for emissivity measurements, emission from a sample cavity is quantified and compared to emission from a blackbody using two types of pyroelectric detectors (typically a liquid nitrogen-cooled mercury cadmium telluride (MCT), or uncooled indium gallium arsenide photodiode (InGaAs). More information regarding the experimental measurement of apparent emissivity using FTIR can be found in the literature [38–40].

An integrating sphere (Fig. 5B) is a hollow chamber with a highly diffuse and reflective interior which collects radiation from a sample cavity or source. The unique spherical shape of the chamber results in a homogenous distribution of the incoming radiation due to multiple reflections.

In integrating sphere reflectometry experiments, the reflectance signal of a sample is compared with that of a reflectance standard whose reflectance is known. Ideally, the ratio of the incoming signals from the sample and the reflectance standard is equal to the ratio of the spectral reflectance of the sample and the reflectance standard. With knowledge of the sample's spectral hemispherical reflectance obtained using this relationship, the normal spectral emissivity for normal incoming radiation to the sample can be obtained using Kirchhoff's law, as discussed in Section 3.3 [41,42].

Even though numerical/computational and experimental methods for determining apparent radiative properties may be preferable in some complex cases, in this work, we have provided a general formulation that can be carried out using any analysis approach (analytical, numerical, or experimental), and show how to use the area integral method and view factors to obtain the apparent behavior of simpler cavities. This approach provides a framework for future analyses with more complex conditions and enables us to communicate how geometry influences heat transfer for isothermal, diffuse, and gray cavities. This understanding can be exploited to control the heat transfer in cavity design.

4. Verification of the general model for apparent radiative surface properties

In this section, we verify the analytical expressions introduced in Section 3 for the apparent emissivity of a generic cavity. To demonstrate the effectiveness of the approach, we have selected an isothermal, diffusely reflecting and emitting cylindrical cavity as a representative case study. The objective is to validate Eq. (19) derived earlier, specifically tailored for these surface conditions. The cylindrical cavity geometry was discretized into

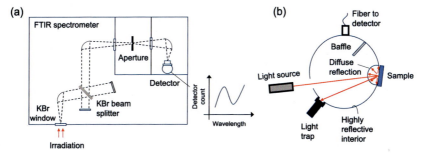

Fig. 5 Commonly used instruments in apparent radiative surface property measurements including (A) a Fourier Transform Infrared Spectrometer (FTIR) and (B) an Integrating Sphere Reflectometer (ISR).

numerous differential elements, allowing the correct quantification of radiation from cavity surfaces to the cavity opening.

In this validation exercise, the analytical expression for the apparent emissivity assumes that the radiosity from different regions of a discretized surface is uniform. It is critical to acknowledge that the view factor and associated radiosity along a cavity surface to the cavity opening may differ at different locations. For instance, in a cylindrical cavity, the opening may receive less irradiation from sidewall regions close to the cavity base, as compared to the middle or upper regions of the sidewall. This variation highlights the non–uniform nature of radiosity which also affects the apparent behavior of the cavity. As a result, subdividing the sidewall and cavity base surfaces will improve the prediction of the total hemispherical apparent emissivity. Therefore, we employ this subdivision of cavity surfaces into numerous differential elements enabling the accurate quantification of irradiation received by the opening from all other surface elements of the cavity and between differential elements. Fig. 6A illustrates the subdivided cylindrical cavity comprised of M differential elements as opposed to treating all cavity walls as a single surface (as shown in Fig. 6B). With an increasing number of elements M, the correct radiosity from different regions of the cavity to the cavity opening will be appropriately addressed and the resulting total hemispherical apparent emissivity can then be compared to results provided in the literature.

The view factors between the sidewall elements and ring-shaped elements of the base were calculated using view factor algebra [25]. The introduction of imaginary surfaces (R) simplifies the determination of view factors from sidewall elements and ring-shaped surfaces on the base. The view factor between the aperture and any imaginary surface (e.g. F_{0-R_1}) was determined by the sum of two view factors: between the aperture and the neighboring side of the imaginary surface (e.g. F_{0-A_2}), as well as the view factor between the aperture and the subsequent imaginary surface (e.g. F_{0-R_2}). This relationship can be expressed for all discretized side wall elements. To determine the view factors between imaginary surfaces, however, we employed a known model for view factors between coaxial parallel disks [25].

Fig. 7 illustrates how the total hemispherical apparent emissivity of an isothermal, diffuse, and gray cylindrical cavity changes with varying numbers of surface elements (M). This variation is depicted for different intrinsic emissivity values and is calculated using Eq. (19). Each plot

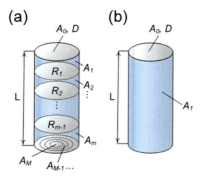

Fig. 6 (A) A discretized cylindrical cavity using a total of M number of elements on the side and base of the cavity. (B) The cylindrical cavity comprised of two surfaces representing the cavity opening (A_0) and all interior cavity surfaces (A_1). To quantify view factors between differential elements of the cavity side and base in case (A), imaginary surfaces (denoted as R_i, where i ranges from 1 to $m-1$) were employed.

represents a different length-to-diameter ratio with L/D ratios of 4 and 1 for Fig. 7A and B, respectively. Dashed lines represent the reported values by Chandos et al. for the apparent emissivity of a cylindrical cavity at the corresponding intrinsic emissivity and a specified L/D ratio [43]. Chandos et al. [43]. applied the same emission and reflection characteristics in computing the apparent emissivity for a cylindrical cavity, although they used a numerical solution of the integral equations. As shown in Fig. 7, we observe the convergence of results from the current work to the results of Chandos et al. as the number of surface elements increases, regardless of the intrinsic emissivity. However, small values of intrinsic emissivity are observed to increase the difference in results between the current work and those reported by Chandos et al. The largest difference between these two works was found to be less than 0.06% for 80 surface elements and less than 1.76% for as few as 16 surface elements. The convergence of these results validates the methodology and approach reported in Section 3. As shown in these results, the analytically derived expressions of this work can be implemented using closed-form expressions or assisted by computational tools, depending on the desired accuracy of the solution and the complexity of the geometry.

Similarly, Fig. 8 shows how the total hemispherical apparent emissivity of an isothermal, diffuse, and gray cylindrical cavity changes with varying numbers of surface elements (M). This variation is depicted for various L/D ratios, with each figure representing a different intrinsic emissivity

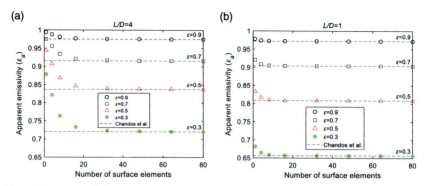

Fig. 7 Variation of the total hemispherical apparent emissivity of an isothermal, diffuse, and gray cylindrical cavity with number of surface elements for (A) $L/D = 4$ and (B) $L/D = 1$. Regardless of the intrinsic emissivity value, a higher number of surface elements representing the cavity results in greater agreement between the approach of this work and the published data by Chandos et al. [43].

value. Intrinsic emissivity values of 0.5 and 0.3 were selected as they were observed to result in the largest difference between the current work and Chandos et al., as noted above. For a small number of surface elements, an increase in the difference between the current work and those reported by Chandos et al. is observed with increasing L/D ratio.

While increasing the number of surface elements allows one to correctly represent radiosity from different regions of the cavity resulting in accurate predictions of the total hemispherical apparent emissivity, relatively few elements are required to obtain reasonable accuracy. As shown in Figs. 7 and 8, a reduced number of surface elements always overpredicts the actual total hemispherical apparent emissivity and can be used as an upper bound. Further, the amount of overprediction reduces with increasing intrinsic emissivity and decreasing L/D ratio.

Fig. 9A illustrates the influence of increasing the number of surface elements and L/D ratio on the relative difference between the total hemispherical apparent emissivity for an isothermal, diffuse, gray cylindrical cavity calculated in this study and the values presented by Chandos et al. [43]. The intrinsic emissivity was set to 0.3 to illustrate the maximum difference between the two works. It is evident that regardless of the L/D ratio, increasing the number of surface elements (i.e., $M \geq 16$) leads to a small relative difference between the results of the two studies. The largest relative difference is observed for a cylindrical cavity characterized by an L/D ratio of 4 and composed of only two surfaces; however, a closed-form expression for the total hemispherical apparent emissivity under these conditions can be readily obtained

Review of the cavity effect

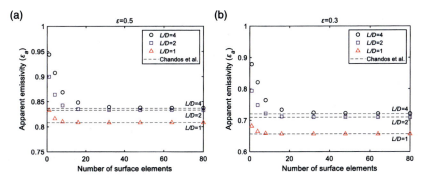

Fig. 8 Variation of the total hemispherical apparent emissivity of an isothermal, diffuse, and gray cylindrical cavity with number of surface elements for intrinsic emissivities of (A) $\varepsilon = 0.5$ and (B) $\varepsilon = 0.3$. Regardless of the length-to-diameter ratio, a higher number of surface elements representing the cavity results in greater agreement between the approach of this work and the published data by Chandos et al. [43].

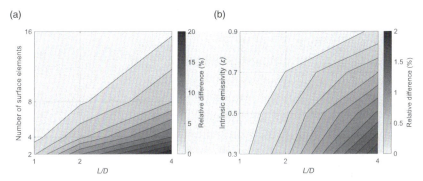

Fig. 9 A 2D contour plot illustrating the relative difference between the total hemispherical apparent emissivity for an isothermal, diffuse, and gray cylindrical cavity obtained in this study, and that reported data by Chandos et al. [43]. for various length-to-diameter ratios (L/D). Plot (A) indicates the relative difference for $\varepsilon = 0.3$ as the number of surface elements varies and plot (B) indicates the relative difference for $M = 16$ surface elements as the intrinsic emissivity (ε) varies.

as will be discussed in Section 5. Regardless, Fig. 9 demonstrates that even a two-surface enclosure approach can lead to reasonable results, especially for shallow cavities or higher intrinsic emissivity, as will be shown.

Fig. 9B demonstrates the relative difference between total hemispherical apparent emissivity for an isothermal, diffuse, gray cylindrical cavity calculated in this study and the data reported by Chandos et al. as L/D ratio and intrinsic emissivity vary. For Fig. 9B, 16 surface elements were used to model the cavity since Figs. 7 and 8 indicate the good agreement expected for this number of surface elements. Overall, the disparity between the approach reported in our study and that of Chandos et al. remains less than 2% across various L/D ratios and intrinsic emissivity values for $M = 16$ surface elements, with an even greater agreement for increasing M (as shown in Figs. 7 and 8).

The largest discrepancy between the results of these two studies is observed at the highest reported L/D ratio and the lowest intrinsic emissivity. In contrast, the smallest difference between the reported data occurs at the smallest L/D ratio, irrespective of the intrinsic emissivity value. It is observed that the impact of increasing the intrinsic emissivity becomes more pronounced at high L/D ratios.

5. Apparent emissivity for isothermal, diffuse, gray, two-surface enclosures

In the preceding section, the validation exercise was limited to a cylindrical geometry for which comparison data was readily available. Using Eq. (19) for an isothermal, diffuse, and gray cavity, one can determine the total hemispherical apparent emissivity for any cavity shape, once view factors are determined between any number of discretized surfaces comprising the cavity. As observed in the comparisons of Figs. 7 and 8, reasonable comparisons of the total hemispherical apparent emissivity can be obtained between different geometries by using even simple two-surface enclosures, with one surface representing an isothermal cavity interior and one surface representing the cavity opening. Such an approach offers a closed-form expression that can be readily used to calculate apparent emissivity.

This section provides the closed-form, analytical solution for the apparent emissivity of six common cavity shapes under the conditions of isothermal, diffuse, and gray cavity surfaces and assuming a two-surface enclosure. The apparent emissivity referred to throughout this section is the total hemispherical emissivity representing the cavity opening. Cavity shapes include cylindrical, conical, spherical, cylindro-conical, cylindro-inner cone, and double-cone cavities. The geometric variables associated with these six cavities are shown in Fig. 10A-F. The apparent emissivity for each cavity shape is obtained by applying Eq. (19) as developed for the specified conditions. Further discussion of how the cavity geometry affects the apparent emissivity is provided in Section 5.7, as the apparent emissivity for each cavity shape is compared to other shapes. In Section 7, we further explore the agreement of our findings for a two-surface enclosure on the apparent behavior of different cavity shapes with other studies that employ numerical, computational, and experimental approaches. Additionally, we provide detailed insights into the key factors that significantly influence the apparent emissivity of various cavity shapes.

5.1 Cylindrical cavity

A cylindrical cavity is one of the most common cavity shapes and is frequently employed in approximating blackbody emission (Fig. 10A). It can be thought of as being comprised of two surfaces, the cylindrical body and the imaginary surface representing the opening. As noted in Section 4, a two-surface approach will overestimate the apparent emissivity but provides a reasonable starting value and offers a readily obtainable expression for calculation. Employing the analysis for isothermal, diffuse and gray surfaces from Eq. (19), the apparent emissivity for a cylindrical cavity is as follows,

$$\varepsilon_a = \frac{F_{01} J_1}{\sigma T^4} \qquad (20)$$

where subscript 0 refers to the cavity opening, and 1 denotes the cylindrical cavity interior surface. Also, emissivity ε and temperature T refer to the interior cylindrical surface intrinsic emissivity and temperature ($T = T_{eff}$), respectively. Since $F_{01} = 1$ and assuming the cavity opening has no emission or reflection ($J_0 = 0$), Eq. (20) becomes:

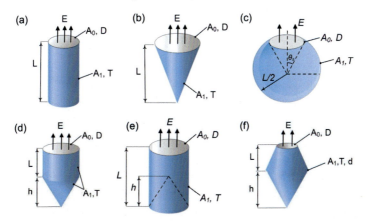

Fig. 10 Schematics illustrating geometric conditions for (A) cylindrical, (B) conical, (C) spherical, (D) cylindro-conical, (E) cylindro-inner cone, and (F) double-cone cavities.

$$\varepsilon_a = \frac{J_1}{\sigma T^4} = \frac{\varepsilon \sigma T^4 + (1 - \varepsilon) F_{11} J_1}{\sigma T^4} \tag{21}$$

By definition, the apparent emissivity leaving surface 1 is the radiosity of surface 1 normalized by the blackbody emissive power evaluated at the cavity surface temperature (as shown above). Rearranging Eq. (21) enables the determination of radiosity leaving the inner walls of the cylindrical cavity (J_1).

$$J_1 = \frac{\varepsilon \sigma T^4}{1 - (1 - \varepsilon) F_{11}} \tag{22}$$

Applying the view factor summation rule and reciprocity relation [25], F_{11} is determined from knowing $F_{01} = 1$.

$$F_{11} = 1 - \frac{A_0}{A_1} F_{01} = 1 - \left(1 + \frac{4L}{D}\right)^{-1} \tag{23}$$

Substituting the view factor F_{11} into J_1, and J_1 into Eq. (21), we obtain the following expression for the apparent emissivity of a cylindrical cavity, consistent with that available in the literature [25].

$$\varepsilon_a = \frac{1}{1 + \left(\frac{1-\varepsilon}{\varepsilon}\right)\left(1 + \frac{4L}{D}\right)^{-1}} \tag{24}$$

5.2 Conical cavity

Determination of the apparent emissivity for each of the six geometries in this work all follow the same analysis approach as that shown above for a cylinder (with subscript 0 referring to the cavity opening, and 1 referring to the isothermal cavity interior surfaces, see Fig. 10). As a result, Eq. (21) and (22) apply for any cavity geometry (including the six considered here), if the interior walls are assumed to be isothermal, diffuse, and gray. The difference in the apparent emissivity among the cavity geometries is due to view factor F_{11}, as specified by the cavity geometry.

For a conical cavity (Fig. 10B), we obtain F_{11} by applying the view factor summation rule and reciprocity relation as before, with $F_{01} = 1$.

$$F_{11} = 1 - \frac{1}{2}\left[\left(\frac{L}{D}\right)^2 + \frac{1}{4}\right]^{-0.5} \tag{25}$$

Substituting F_{11} into Eq. (22), we obtain the apparent emissivity for a conical cavity [25].

$$\varepsilon_a = \cfrac{1}{1 + \left(\frac{1-\varepsilon}{2\varepsilon}\right)\left[\left(\frac{L}{D}\right)^2 + \frac{1}{4}\right]^{-0.5}}$$

(26)

5.3 Spherical cavity

Fig. 10C illustrates a spherical cavity with a circular opening. The apparent emissivity for the spherical cavity can be either developed as a function of the cavity opening angle or as a function of the cavity diameter. Although both formulations are expressed here, the latter is used to compare the apparent emissivity of a spherical cavity with other cavity shapes.

Considering the cavity surface comprised of the cavity opening and sphere, the relation for F_{11} is obtained,

$$F_{11} = 1 - \left[4\left(\frac{L}{D}\right)^2 - 1\right]^{-1}$$

(27)

where L is the sphere diameter and D is the opening diameter. Substituting F_{11} into Eq. (22), we obtain the apparent emissivity for a spherical cavity [25,46].

$$\varepsilon_a = \cfrac{1}{1 + \left(\frac{1-\varepsilon}{\varepsilon}\right)\left[4\left(\frac{L}{D}\right)^2 - 1\right]^{-1}} = \cfrac{1}{\frac{1}{\varepsilon} - \frac{(1-\varepsilon)}{2\varepsilon}(1 + \cos\theta_s)}$$

(28)

Alternatively, the apparent emissivity can be determined by defining the cavity opening in terms of the angle corresponding to the cavity radius (Fig. 10C). For the opening $A_0 = \pi(R\sin\theta_s)^2$ and cavity walls $A_1 = 2\pi R^2[1 + \cos\theta_s]$, the view factor F_{11} is determined.

$$F_{11} = \frac{1}{2}[1 + \cos\theta_s]$$

(29)

Substituting F_{11} into Eq. (22), we obtain the apparent emissivity for a spherical cavity in terms of the angle of the cavity opening (right side of Eq. (28)). Note that as θ_s approaches 0, F_{11} approaches unity as does the apparent emissivity of the spherical cavity.

5.4 Cylindro-conical cavity

Fig. 10D illustrates a cylindro-conical cavity comprised of a cylindrical body with a conical bottom. Treating the conical base and cylindrical sidewalls as a single surface, the view factor F_{11} associated with this shape is obtained as follows.

$$F_{11} = 1 - \left(\frac{4L}{D} + 2\left[\left(\frac{h}{D} \right)^2 + \frac{1}{4} \right]^{0.5} \right)^{-1} \tag{30}$$

Substituting F_{11} into Eq. (22), we obtain the apparent emissivity for the cylindro-conical cavity.

$$\varepsilon_a = \frac{1}{1 + \left(\frac{1-\varepsilon}{\varepsilon} \right) \left(\frac{4L}{D} + 2\left[\left(\frac{h}{D} \right)^2 + \frac{1}{4} \right]^{0.5} \right)^{-1}} \tag{31}$$

5.5 Cylindro-inner cone cavity

The cylindro-inner cone cavity is comprised of a cylinder with a reentrant cone, as shown in Fig. 10E. The view factor F_{11} associated with this cavity based on the surface area of the body and imaginary circular opening is as follows.

$$F_{11} = 1 - \left(\frac{4L}{D} - 2\left[\left(\frac{h}{D} \right)^2 + \frac{1}{4} \right]^{0.5} + 2 \right)^{-1} \tag{32}$$

Therefore, the apparent emissivity of the cylindro-inner cone cavity is,

$$\varepsilon_a = \frac{1}{1 + \left(\frac{1-\varepsilon}{\varepsilon} \right) \left(\frac{4L}{D} - 2\left[\left(\frac{h}{D} \right)^2 + \frac{1}{4} \right]^{0.5} + 2 \right)^{-1}} \tag{33}$$

5.6 Double-cone cavity

A generalization of a cylindro-conical cavity is a double-cone (Fig. 10F) consisting of a conical base combined with a conical frustum terminating in an annular opening. Again, treating conical and frustum sections as single surface, the view factor is,

$$F_{11} = 1 - \frac{1}{2} \left\{ \left(\frac{d}{D} \right)^2 \left[\left(\frac{h}{d} \right)^2 + \frac{1}{4} \right]^{0.5} + \frac{(d + D)}{D} \right.$$
$$\left. \left[\frac{1}{4} \left(\frac{d}{D} \right)^2 - \frac{1}{2} \left(\frac{d}{D} \right) + \frac{1}{4} + \left(\frac{L}{D} \right)^2 \right]^{0.5} \right\}^{-1} \tag{34}$$

The apparent emissivity of the double-cone cavity is then obtained as follows.

$$\varepsilon_a = \cfrac{1}{1 + \left(\frac{1-\varepsilon}{2\varepsilon}\right)\left\{\left(\frac{d}{D}\right)^2\left[\left(\frac{h}{d}\right)^2 + \frac{1}{4}\right]^{0.5} + \frac{(d+D)}{D}\left[\frac{1}{4}\left(\frac{d}{D}\right)^2 - \frac{1}{2}\left(\frac{d}{D}\right) + \frac{1}{4} + \left(\frac{L}{D}\right)^2\right]^{0.5}\right\}^{-1}}$$

(35)

In Section 5.7, analytical results of the apparent emissivity assuming a two-surface enclosure for the six cavities are reported and compared. The influence of varying geometric and material properties of cavity surfaces on the apparent emissivity is also explored.

5.7 Geometric comparison

The apparent emissivity for each of the six geometries presented earlier (cylindrical, spherical, conical, cylindro-conical, cylindro-inner cone, and double-cone cavities) are presented and compared in Fig. 11. The effects of the following conditions are explored: (1) intrinsic emissivity (ε); (2) length-to-diameter ratio (L/D); (3) cavity diameter to the opening ratio (i.e., L/D for spherical cavity only); and (4) conical height-to-diameter ratio (i.e., h/D for cylindro-conical, cylindro-inner cone and double-cone cavities only).

Fig. 11 reports the total hemispherical apparent emissivity as a function of L/D ratio for six cavity geometries. As expected, an increase in the intrinsic emissivity ε and L/D ratio increases the apparent emissivity, regardless of cavity shape. The impact of changing L/D is even more noticeable for lower intrinsic emissivities. Conversely, for the highest intrinsic emissivity ($\varepsilon = 0.9$), changing the L/D only slightly affects the apparent emissivity regardless of cavity shape, with the smallest effect observed in the double-cone (Fig. 11F), spherical (Fig. 11C), and cylindro-inner cone cavities (Fig. 11E). Furthermore, it is observed that for $\varepsilon = 0.9$, the apparent emissivity is greater than 0.9, approaching that of a blackbody, as L/D approaches 10 for any cavity shape. For small L/D values this cavity effect diminishes, and the apparent emissivity approaches the intrinsic emissivity of cavities.

In light of Fig. 11, it can be observed that the commonly used cylindrical and conical cavities have lower apparent emissivities among all cavity shapes for a given condition (i.e., equivalent intrinsic emissivity, surface properties, etc.). The spherical and double-cone cavities demonstrate the highest average apparent emissivity among all cavity shapes though, followed by cylindro-conical, cylindro-inner cone, cylindrical and conical cavities. It is likely that, although the

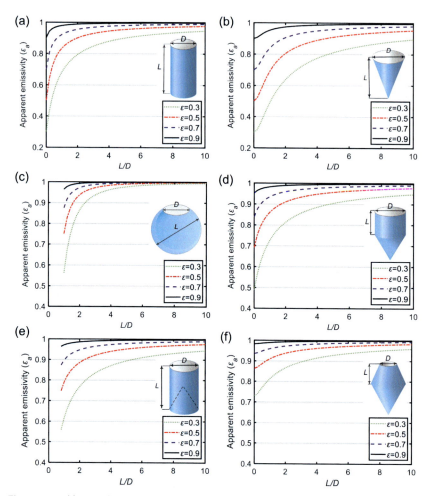

Fig. 11 Total hemispherical apparent emissivity for isothermal, diffuse, and gray interior cavity surfaces for the following cavity geometries: (A) cylinder, (B) cone, (C) sphere (D) cylindro-cone, (E) cylindro-inner cone, and (F) double-cone. Results are shown for varying length-to-diameter ratios (L/D) and intrinsic surface emissivity (ε) and assume a two-surface enclosure. For the sake of comparison, the height-to-diameter (h/D) ratio for cylindro-conical, cylindro-inner cone and double-cone cavities was set to 1. The conical-to-frustum opening ratio (d/D) for the double-cone cavity was set to 1.5.

cylindrical and conical shapes exhibit lower apparent emissivity for the same conditions, these geometries have been favored due to their simplicity of construction. Fig. 11 also indicates that deeper cavities correspond to higher apparent emissivities. A deeper cavity can be achieved by increasing either L/D or h/D (for cylindro-conical, cylindro-inner cone, and double-cone cavities).

The influence of h/D on the apparent emissivity for the cylindro-conical, cylindro-inner cone and double-cone cavities with $\varepsilon = 0.7$ is illustrated in Fig. 12. To obtain results for the double-cone cavity, the ratio of the conical diameter to the frustum opening (d/D) was set to 1.5. It is observed that increasing h/D results in an increase in the apparent emissivity of cylindro-cone and double-cone cavities. However, the cylindro-inner cone cavity shows an entirely opposite behavior. Increasing h/D for a cylindro-inner cone (i.e., decreasing the depth of the cavity) reduces the apparent emissivity. However, decreasing h/D does not always ensure higher apparent emissivity, as the length of the cylindrical section can have a more significant influence on the apparent emissivity. Additional consideration should be given to determine the optimum depth, height, and diameter of a cylindro-inner cone cavity, targeted to a specific application. Nevertheless, at high L/D ratios, h/D hardly affects the apparent emissivity and all three cavity shapes exhibit near blackbody behavior having the apparent emissivity close to unity. Also shown in Fig. 12 is that the double-cone cavity has a relatively higher average apparent emissivity among these three cavity shapes, implying greater blackbody-like behavior.

A comparison of the apparent emissivity of all six cavity geometries with respect to L/D (assuming isothermal, gray, and diffuse cavity surfaces) is depicted in Fig. 13 ($\varepsilon = 0.7$ for all cases). For cylindro-conical and cylindro-inner cone cavities $h/D = 1$; for the double-cone cavity $h/D = 1$ and $d/D = 1.5$. As mentioned earlier, at higher L/D ratios, the spherical cavity exhibits the highest apparent emissivity among all cavities. This means, for a spherical cavity, a larger L/D ratio results in a smaller cavity opening to the sphere diameter ratio. The result for this case is consistent with the published data in the literature for the radiative heat loss from a spherical cavity [47].

Moreover, the double-cone cavity shows the highest apparent emissivity for $L/D < 2$ under the same surface conditions and material/geometric properties. As the cavity becomes deeper, the apparent emissivity converges to that of a blackbody cavity. A more drastic increase is seen in the apparent emissivity of the double-cone and conical cavities with increasing L/D ratios when compared to other cavity shapes.

Throughout Section 5, we reported analytical expressions for the apparent emissivity of different cavity shapes with isothermal, diffuse, and gray walls. Further, we quantified the effect of varying geometric parameters on the apparent emissivity. Validation of modeling approaches in the literature often includes spectrometry, integrating sphere reflectometry, and other experimental approaches. These measurement methods can be

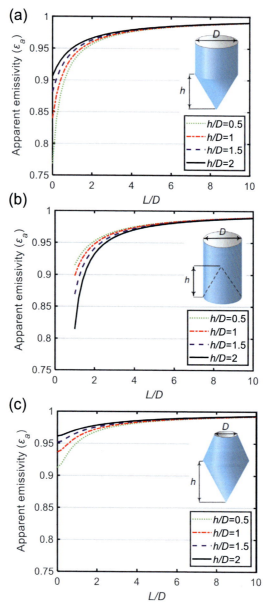

Fig. 12 Total hemispherical apparent emissivity for isothermal, diffuse, and gray interior cavity surfaces for the following cavity geometries: (A) cylindro-cone, (B) cylindro-inner cone, and (C) double-cone. Results are shown for varying length-to-diameter ratio (L/D) and conical height-to-diameter ratio (h/D) and assume a two-surface enclosure. The intrinsic emissivity was set to 0.7 for all plots. The conical to the frustum opening ratio (d/D) for the double-cone cavity was set to 1.5.

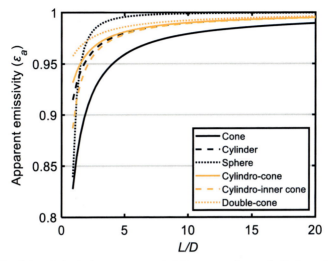

Fig. 13 Total hemispherical apparent emissivity comparison of all six cavity shapes (cylinder, cone, sphere, cylindro-cone, cylindro-inner cone, and double-cone) with respect to the length-to-diameter ratio (L/D) and assuming a two-surface, isothermal enclosure. The intrinsic emissivity was set to 0.7 for all cavities. For cylindro-conical and cylindro-inner cone cavities $h/D = 1$; for the double-cone cavity $h/D = 1$ and $d/D = 1.5$.

employed to determine the apparent emissivity for various cavity shapes and study the influence of varying surface conditions, geometric properties, and cavity heating mechanisms [40,48–52]. In the following section, published works associated with three forms of apparent emissivity described in Section 3.4 are summarized. The key focus of the following section is to outline the advances in cavity modeling and assess the influence of altering the geometric properties of cavities on the apparent radiative properties of different cavity shapes.

6. Advances in cavity modeling

The cavity effect for different cavity shapes has been widely discussed by researchers over the last century [5,29,30,34,53–57]. The complexity of evaluating the radiative behavior of these cavities and their dependence on geometric properties, material properties, temperature, and wavelength has led to ongoing investigation. In this section, cavity analyses performed by researchers in the literature are summarized and discussed with the primary focus on identifying critical factors affecting apparent radiative surface properties for a given cavity shape. Here, we organize the literature by cavity shape and then additionally categorize the literature according to the following conditions: (1) *diffusely* reflecting walls, or (2) walls that are either *specular* (mirror-like) or *diffuse-specular*. For this section, the emission is considered as diffuse only, and the surfaces are opaque.

The purpose of these sections is to identify the influence of geometric control on the apparent radiative surface properties and critical parameters that would enable one to dynamically control radiative properties. As a result, the modification of apparent radiative surface properties can then act as a means of temperature regulation. Geometries are introduced in the same order as that for the sections above, namely: cylindrical, spherical, conical, cylindro-conical, cylindro-inner cone, and double-cone cavities.

6.1 Cylindrical cavity

6.1.1 Diffusely reflecting and emitting surfaces

Surfaces may be idealized as diffuse or specular based on the manner in which they reflect irradiation. Diffuse reflection occurs if, irrespective of the direction of incident irradiation, the intensity of reflected energy is the same in all directions [25]. Excellent discussions regarding the behavior of diffuse cylindrical cavities exist in the literature. We begin with early analytical models that were proposed for radiative heat transfer in cylindrical cavities, clearly identifying all the approximations, methods, and limitations of these approaches for determining apparent radiative properties. Next, we summarize numerical modeling efforts and the most recent works for apparent radiative properties computations. Some notable examples of improving the cylindrical cavity shape for better radiative performance are also highlighted at the end of this section.

Radiation exchange in an isothermal cylindrical cavity was first introduced by Buckley [58]. It has been claimed that radiation exchange inside a cylindrical cavity is governed by one or more nested integral equations (when accounting for the cavity wall and the base separately) [20,59–61]. Although different researchers began with the same governing equations, some simpler approaches have been employed with various approximations to solve these equations. Gouffe [62] assumed a uniform distribution of reflected intensity inside the cavity and determined the fraction ($\rho\Omega/\pi$) of irradiation reflected after the first reflection in terms of the opening surface and cavity wall areas. All subsequent reflections then were estimated as a function of the first reflection to derive an expression for the apparent emissivity of different cavity shapes, including the cylinder. Further, Gouffe stated that only the first two reflections had a significant impact on the apparent emissivity in this work.

Likewise, Treuenfels [63] attempted to quantify the apparent emissivity of different cavities including a cylinder accounting for successive reflections inside the cavity and employing Kirchhoff's law. It was stated by Treuenfels that if f_n is the fraction of rays reflected n times, then the apparent reflectivity can be determined using $\rho_a = \Sigma f_n \rho^n$ where $f_n = f_1 (1 - f_1)^{n-1}$, ρ is the reflectance of cavity walls and f_1 is the fraction of first reflectance. Then, one can easily determine the apparent emissivity for a cavity comprised of opaque and diffuse walls using Kirchhoff's law ($\varepsilon_a = 1 - \rho_a$). Treuenfels used an angle factor that was averaged over the entire cavity surface, which resulted in the hemispherical quantity of the apparent emissivity. Similar to Gouffe's work, the core issue in this approach was to compute the fraction of the first reflection (f_1) of the incident irradiation. This fraction has been computed for a variety of cavities using: (1) the fundamental definition of a view factor for diffuse reflection as carried out by Treuenfels; and (2) radiation modeling using two imaginary ring elements involved in the radiation exchange inside the cavity when the directional variation of apparent emissivity is favored [64]. Results from Treuenfels' work using approach (1) for the variation in apparent emissivity as a function of intrinsic emissivity for a cylindrical cavity (and for a conical cavity) were in disagreement with data reported by Gouffe [62].

However, results in these early works for the apparent emissivity generally showed good agreement only for cavity walls with high intrinsic emissivity [65]. Some researchers have attempted to reduce the number of approximations in deriving the integral equations by considering a realistic reflection behavior and temperature distribution inside the cavity [66,67]. In particular, efforts were made to replace the view factors used in the integral equations with approximate exponential functions to derive an analytical expression for the apparent emissivity of a cylindrical cavity [68,69]. It has been observed that in a diffuse cylindrical cavity with $\varepsilon = 0.5$, a quadratic temperature profile as a function of L/D increases the integrated total apparent emissivity from 0.9350 for an isothermal cavity to 0.9632 [70–72].

In early works, results for the apparent emissivity have normally been presented by solving a single integral equation for the hemispherical apparent emissivity [53]. However, Sparrow et al. expressed the 'local' apparent emissivity using two nested integral equations accounting for both radial and axial variations [68,70]. They solved these nested integral equations using an iterative trial-and-error procedure.

Integral equations representing the radiative behavior of the cavity can be solved through a variety of other techniques. Approximation of the integral equations solution with infinite series using zonal approximation has been employed by some researchers [73]. In this method, one can divide the cavity's internal surface into N bands (in addition to the geometric and local cavity property variations) and assess the cavity effect with respect to the bandwidth of radiation [74]. The influence of linear and quadratic temperature profiles for cavity walls on the apparent emissivity can also be evaluated with ease [75]. Application of various analytical methods, simplifying assumptions, cavity material or geometric properties, as well as detector conditions (i.e., detector size and distance between cavity and detector), have also been investigated in the literature [68,71,76]. One common observation among all works is that the apparent emissivity for a cylindrical cavity experiences an increase as the length-to-diameter ratio (L/D) and intrinsic emissivity increase (ε). For example, it has been shown that for $\varepsilon = 0.9$ and $L/D = 4$ the integrated and hemispherical apparent emissivities reach 0.994 and 0.974, respectively, providing nearly black behavior [77]. Also, it has been shown that, unlike the distance between the cavity and the detector, the size of the detector may not have a significant impact on integrated normal emissivity [76].

As a result of the increasing prevalence of computational tools and advanced integration techniques, the problem of solving nested integral equations is no longer as problematic. MCRT is one numerical method that is commonly used to obtain local, integrated, and hemispherical apparent emissivities for various cavity shapes when closed-form solutions do not exist or are hard to obtain due to the complexity of the cavity geometry or propagation of incident and reflected radiation [43]. MCRT simulates emission from complex cavity shapes with ease and has been used to quantify the apparent radiative behavior of isothermal/non-isothermal cylindrical cavities with greater rigor. As mentioned in Section 3.5, absorption and emission schemes of the MCRT method have proven to produce accurate and efficient results when dealing with complex geometries. MCRT does not normally result in a divergence from the correct solution when dealing with complex geometries and can be employed with less computational time as compared to other numerical methods [32]. MCRT can also enable additional geometric complexity, such as the addition of grooves within the cylindrical cavity and the ability to compute the angle factor between two surface elements [78]. As one example, for an opaque, diffuse, cylindrical cavity with an $L/D = 10$, the local apparent emissivity at the cavity base after ten reflections becomes 0.999 using the hit-or-miss MCRT method for 10^5 rays entering aligned with the axis of the cavity [79].

In addition to the implementation of various analysis approaches, there have been many efforts to modify the cylindrical cavity geometry and thereby enhance the net radiative emission of a cylindrical cavity, often with the objective of increasing the apparent emissivity and obtaining a near-blackbody emitter [80]. Grooved cylindrical cavities at the base (Fig. 14) have been shown to have a more uniform apparent emissivity as compared to a cylinder with a flat bottom [29].

Applying circular baffles at the opening and applying grooves to the cylindrical cavity base to reduce first and second successive reflections also provide an increase in the apparent emissivity [81]. It has become increasingly common to add an annular cavity aperture to enhance the cavity effect [82]. For isothermal conditions, this aperture addition results in one additional integral equation for a total of three: one for the cylindrical body, base, and aperture of the cavity. The closed-form solution of a lidded cavity has not been reported in the literature to date, but numerical integration has been performed to solve the three integral equations simultaneously [82]. For $\varepsilon = 0.5$, $L/D = 4$ and a dimensionless diaphragm radius of 0.4 (ratio of the diaphragm radius to cavity radius), adding a diaphragm increases the apparent emissivity of an isothermal and diffuse cylindrical cavity from 0.8389 (lidless case) to 0.9679. It should be noted that in determining the apparent emissivity, it is normally assumed that cavity walls are insulated. This assumption simplifies the solution procedure since 2D conduction heat transfer with surroundings is neglected. However, there are many situations in which the boundary temperature is unknown, and the effect of conduction heat transfer is not negligible [83]. Observations show that taking conduction heat transfer

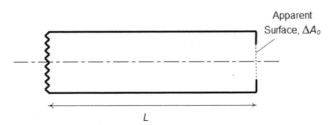

Fig. 14 A cross-sectional schematic of a cylindrical cavity, featuring baffles at the opening and a solid, grooved base.

into account in the radiation heat exchange between a cylindrical cavity and its surroundings and assuming a 2D temperature distribution for the cavity walls can have a significant impact on the radiative flux streaming out of the cavity [84,85].

6.1.2 Specularly reflecting and diffusely emitting surfaces

The reflection of incident radiation can be either diffuse, specular, or diffuse-specular (i.e., partially specular [24]). Diffuse-specular reflectivity is commonly defined using the expression $\rho = \rho^d + \rho^s$, where ρ^d denotes the diffuse component of reflection and ρ^s refers to the specular component of reflection. The ratio of ρ^s/ρ is called the specularity ratio and ρ^d/ρ is also called diffusity [2]. Specular or mirror-like surfaces reflect irradiation at the same angle to the surface normal as the incident ray, but on the opposing side of the surface normal and in the same plane formed by the incident and reflected rays [25]. Although the assumption of diffuse reflection may be reasonable in many engineering applications, some surfaces clearly exhibit specular behavior in practice.

De Vos' theorem was one of the earliest to address a cylindrical blackbody cavity's apparent behavior using specular reflection [86]. This theorem is based on the fact that diffuse incident irradiation after multiple reflections inside a specular cavity is no longer diffuse. For simplicity, early works (including De Vos' work) only accounted for the first and second-successive reflections (called first and second-order approximations) in solving the integral equations to obtain apparent radiative surface properties. In more recent works, however, the concept of an 'exchange factor' has been introduced to determine the apparent behavior of curved diffuse-specular cavities. The exchange factor determines the fraction of irradiated energy that leaves an element and arrives at another element both directly and by all possible specular reflections from other elements [87]. Using this approach, it has been shown that the local and overall heat transfer rate leaving a cylindrical cavity (for diffuse-specular surfaces) [88] and the apparent absorptivity of a cylindrical cavity [89] are higher than that of a diffuse cylindrical cavity. Further, the heat transfer rate and apparent absorptivity increase with increasing the intrinsic emissivity of the cavity [87].

For cylindrical cavities with open ends, however, specularly reflecting walls have been shown to increase heat loss from the cavity [90]. The integral equations for this case have been solved by dividing the cavity surface into N + 1 points along the length of the cavity, applying the governing integral equation at each point, and solving the N + 1 linear algebraic equation simultaneously to compute the apparent emissivity of the cavity. For the same case of a *long* cylindrical cavity with open ends (requiring numerous increments N), an alternative method using the Taylor series solution has been employed [90].

Analytical solutions are difficult to obtain for apparent radiative properties especially when dealing with the random nature of reflection inside the cavity. Initially, some researchers believed that analytical expressions for cavities could only be applied when incident irradiation is perpendicular to the apparent surface normal. One study refuted the condition that the irradiation must be perpendicular to the apparent surface normal by building a reflectance measurement apparatus [91]. In the experiment, using an adjustable brass plunger to vary the L/D ratio of the cavity, both diffuse and specular reflections in a cylindrical cavity were measured and compared with Gouffe's data.

As a result of computational advances, many researchers now utilize the MCRT numerical method to quantify apparent emissivity for regular or irregular cavities comprised of diffuse or specular surfaces [92]. For regular curved surfaces (such as cylindrical cavities), using a circle vector function in MCRT, building a new coordinate system, and obtaining reflection functions according to a new coordinate system remarkably reduce the computational time in apparent emissivity calculations [93].

Similar to diffusely reflecting cylindrical cavities, variations in cavity geometry (e.g., L/D ratio) and material properties can greatly impact the apparent radiative behavior of a cylindrical cavity with specularly reflecting walls. Specular cavities have been shown to exhibit relatively higher apparent emissivity than diffusely reflecting cavities, especially at lower intrinsic emissivities when a cavity is highly reflective [94,95]. From the published data in the literature, it is generally perceived that specularly reflecting cylindrical cavities behave closer to a blackbody, but their temperature is less uniform as compared to other cavity shapes [96].

Efforts have been made to further improve the apparent radiative properties of specularly reflecting cylindrical cavities. A specularly reflecting lid (aperture) on top of a cylindrical cavity causes an increase in the normal and hemispherical apparent emissivities. Further, a smaller aperture in the lid and a higher intrinsic emissivity of the cavity walls result in a higher normal apparent emissivity, regardless of the radiative characteristic of the lid [97]. Additional work has been performed to optimize the geometry of cylindrical cavities with specularly reflecting walls to enhance net radiative heat transfer. To increase the number of reflections inside a cavity, an isothermal [2] and a non-isothermal [98] cylinder with an inclined bottom (Fig. 15) have been suggested and evaluated. Using the backward ray tracing method, for the case of an isothermal cavity, it has been shown that the average normal apparent emissivity (when the detector is placed at an infinitely long distance from the cavity opening) substantially increases for inclined bottom angles of 30° or 60°. At an inclined bottom angle of 45°, the average normal apparent emissivity reaches a minimum because the diffuse component of reflection grows remarkably [2].

For a cylindrical cavity with an aperture and inclined bottom but with non-isothermal walls, it has been found that spectral directional apparent emissivity becomes more uniform with increasing wavelength. Further, the point of isothermality (i.e., the depth below which the cavity is isothermal) strongly affects the apparent emissivity. The larger the isothermal region, the higher and more uniform the apparent emissivity [99]. A more specular surface results in an increase in the apparent emissivity, especially at inclined bottom angles close to 60° [98].

A semi-circular cylindrical cavity has also been suggested to enhance radiative properties as compared to a complete cylindrical cavity. The effects of oblique and parallel incident radiation on the apparent absorption of this type of cavity have been investigated. It has been perceived that at oblique angles of incident irradiation, the diffuse semi-circular cavity has a relatively higher apparent absorptivity (emissivity) than that of a specular case due to a shadowing effect [100].

6.2 Conical cavity
6.2.1 Diffusely reflecting and emitting surfaces
Like cylindrical cavities, integral equations have been used to determine the apparent behavior of isothermal or non-isothermal conical cavities and temperature distribution within the cavity [101]. These integral equations have been solved numerically using approximate expressions for view

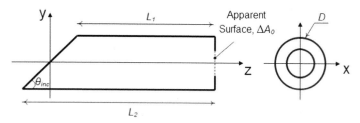

Fig. 15 Cross section of a lidded cylindrical cavity with an inclined bottom.

factors [102]. They have also been approximated using a power series for a conical cavity surface divided into N bands where each band has a constant apparent emissivity [103]. The view factor and local/overall heat transfer rate leaving the conical cavity have also been analytically derived by analyzing the heat exchange between two infinitesimal ring elements inside the cavity [56].

The apparent emissivity of conical cavities is normally evaluated with respect to the change in the cone angle (θ_c) or length-to-diameter ratio (L/D), and the intrinsic emissivity (ε) [87,104]. It has been shown that the apparent emissivity of a conical cavity increases with decreasing cone angle (or increasing L/D) and increasing intrinsic emissivity. The local heat transfer is minimum at the cone apex and increases at this location with increasing the opening angle and intrinsic emissivity of the cavity [88,104].

MCRT and the probability distribution of ray bundles have also been used for analyzing conical cavities [105–107]. Using this approach, it has been shown that small opening angles yield strong reflection toward the cavity walls, while results for large angles approach those for a flat plate [105], as expected.

Efforts to improve the apparent emissivity of a conical cavity include adding an annular baffle or aperture (at the same temperature as the cavity wall) that partially restricts the cavity opening [108]. The use of this annular baffle drastically increases the apparent emissivity of a conical cavity and with an even greater effect on cavity walls with lower intrinsic emissivity. Furthermore, the combination of simple cavity shapes like a cone and a frustum to enhance emission from conical cavities has been suggested by some researchers [55]. For instance, the integrated emissivity for a frustum-inner-cone was observed to reach as much as 0.999 when the distance between the cavity opening and the detector is sufficiently long.

6.2.2 Specularly reflecting and diffusely emitting surfaces

To achieve near-blackbody behavior, a special type of specularly reflecting conical cavity having an inclined bottom (similar to Fig. 15) has been proposed and evaluated [109]. For the case of a long cavity (small opening angle), high intrinsic emissivity ($\varepsilon = 0.9$), and a 30° cone angle, the apparent emissivity was shown to reach 0.999. This type of conical cavity with an inclined bottom has been proposed with a specific application in Cryogenic Solar Absolute Radiometer (CSAR) [109].

6.3 Spherical cavity

6.3.1 Diffusely reflecting and emitting surfaces

Due to the near blackbody behavior of isothermal spherical cavities, several reported works to formulate the apparent emissivity of this cavity shape exist in the literature. Gouffe attempted to quantify the normal apparent emissivity for a spherical cavity, accounting for only the first two reflections inside the cavity [62]. In deriving the normal apparent emissivity, due to the unique geometry of the sphere, the view factor between discrete surface elements was assumed to be constant resulting in a uniform apparent emissivity distribution [62,110,111]. Non-isothermal spherical cases have also been of interest to many researchers. For example, it has been reported that an axial temperature gradient along a spherical cavity can increase the emitted spectral intensity leaving the cavity opening by up to 15 times the Planck distribution at the cavity opening [112].

6.3.2 Specularly reflecting and diffusely emitting surfaces

Similar to other cavity shapes, the assumption of having specularly reflecting surfaces in spherical cavities has also been explored. De Vos [87] obtained an exact expression for the apparent emissivity of a spherical cavity accounting for the actual nature of the reflection [86]. DeVos considered a surface element on the cavity wall (ds) and an imaginary surface element on the cavity opening (da) and simulated all possible reflections from ds and other surface elements reaching da directly or after reflection from ds. One assumption used in this modeling approach was the constant partial reflection of radiation from each surface element incident on the cavity opening in the direction of the unit solid angle of the surface element. Campanaro and Ricolfi [113], however, stated that assuming a constant partial reflection within the solid angle for heat exchange between two surface elements leads to an error in computing the normal apparent emissivity, especially when dealing with surfaces with low absorptivity [114].

For spherical cavities with some degree of specularity (i.e., the diffuse-specular surface condition), the apparent absorptivity has shown a strong dependence on the magnitude of intrinsic absorption of cavity walls [113]. It has been observed that for a spherical cavity having an absorptivity of greater than 0.5, the apparent absorptivity of a diffuse-specular case is higher than that of a perfectly diffuse case for an opening angle between $60°$ and $90°$. In other words, by increasing the intrinsic surface absorptivity above 0.5, the apparent absorptivity of a diffusely reflecting spherical cavity is lower than the equivalent specularly reflecting case. However, for small absorptivities (specifically less than or equal to 0.1), a diffuse spherical cavity always has an equal or higher apparent absorptivity than that of an equivalent specular or diffuse-specular case. Further, regardless of the nature of the reflection, increasing the cavity opening angle always results in a reduction in the apparent absorptivity [113]. A more recent paper examining diffuse-specular behavior in a spherical cavity (which introduced a novel coordinate system to conveniently set up radiation heat transfer integral equations) showed that, unlike cylindrical and conical cavities, the apparent absorptivity does not always increase with increasing specularity ratio and, therefore, increasing specularity ratio does not guarantee higher apparent absorptivity [115].

6.4 Cylindro-conical cavity

6.4.1 Diffusely reflecting and emitting surfaces

Early work carried out by Bedford et al. [57]. formulated the local apparent emissivity of a cylindro-conical cavity through defining three integral equations, one for each surface comprising the cavity enclosure: conical sidewalls, cylindrical sidewalls, and cavity opening. In their work, the surface of each section of the cavity was divided into infinitesimally small elements where the apparent emissivity could be assumed uniform. By modeling the view factor between two ring-shaped elements involved in the heat exchange and applying a single differentiation to the view factor expression between these elements, the integral equations were simplified to a set of algebraic equations that could be solved simultaneously. The integral equations for this type of cavity have also been solved using numerical methods, such as Gauss Legendre Quadrature (GLQ), which is a form of Gaussian quadrature for replacing the definite integrals with sums, leading to linear algebraic equations [116]. The MCRT method has also been employed to quantify the uncertainty associated with local apparent emissivity in a diffusely reflecting cylindro-conical cavity. It has been shown that by increasing the conical section depth, a remarkable drop in the local apparent emissivity of the cylindrical section is observed while exhibiting a more uniform local apparent emissivity in the conical bottom [117].

For a cylindro-conical cavity, the local apparent emissivity is maximum at the conical section vertex, and it reduces along the conical section towards the cavity opening. In thermometry, the portion of the cavity seen from a detector has a remarkable impact on the local apparent emissivity. It has been shown that the common belief of using a 120° cone angle to achieve the highest local apparent emissivity has no rationale and the maximum emission is obtained for a cone angle of around 160°. The local apparent emissivity corresponding to this angle can further be enhanced by increasing ε and L/D. However, reducing the cavity opening diameter (adding a lid/aperture) has a negligible impact on the local apparent emissivity of this type of cavity [116].

6.4.2 Specularly reflecting and diffusely emitting surfaces

Using the Monte Carlo Ray Tracing method, the local and hemispherical apparent emissivities have been evaluated with respect to the change to the cone apex angle (θ_c), specularity ratio (ρ^s/ρ), ε, and cavity L/D for the following cases: specularly-reflecting and diffuse-specular cylindro-conical cavities with one conical bottom and one annular opening [118–120] and for two conical bottoms (one conical and one frustum-shaped bottom) [121–123]. It was observed that for normal incident radiation, the hemispherical apparent emissivity for both cases mentioned above has a local minimum for cone angles at 60° and 90° [119,122]. The reason for this reduction lies in the smaller number of reflections of incident irradiation at these angles. Furthermore, increasing ρ^s/ρ, ε, θ_c, and L/D was shown to increase the hemispherical apparent emissivity.

To further augment the apparent behavior of a cylindro-conical cavity, the influence of rectangular and triangular grooves on the cylindro-conical cavity walls has been investigated for diffusely [124] or specularly reflecting [125] walls and a 120° conical section angle. Generally, the apparent emissivity of a grooved cylindro-conical cavity is higher than that of the smooth case, especially at high specularity ratios and lower intrinsic emissivities. However, the addition of grooves does not necessarily increase the apparent emissivity as the specularity ratio grows. This is because the fraction of incident radiation leaving the cavity after n reflections does not always increase with increasing specularity ratio for grooved cylindro-conical cavities. Triangular grooves, however, result in greater blackbody behavior than the rectangular grooved case, especially at moderate degrees of specularity [125].

In early works to compute apparent radiative properties, cavity surfaces were assumed to have uniform specular or diffuse reflection characteristics over the hemisphere. This assumption neglects the angular dependence of specular reflection as the cavity is considered to have either perfectly diffuse or perfectly specular reflecting surfaces. Even though the uniform diffuse-specular reflection is a realistic assumption used in cavity analysis, the angular dependence of the incoming radiation on the cavity should not be disregarded. To address this, a three-component bidirectional reflectance distribution function (3C BRDF) accounting for diffuse, specular, and glossy components of reflection was proposed. Using this modeling approach, the apparent emissivity of a non-isothermal cylindro-conical cavity was computed and compared with an isothermal case [1]. Results indicated that regardless of the nature of reflection, the isothermal cylindro-conical cavity has a higher integrated apparent emissivity than the non-isothermal case.

The influence of temperature distribution on the apparent emissivity and the location of isothermality inside the cylindro-conical cavity has also been explored by some researchers [126,127]. The variation of integrated apparent emissivity of a cylindro-conical cavity using six types of arbitrary temperature distributions has been studied by Prokhorov et al. [128]. It was observed that the portion of the cavity that is interrogated during measurement can significantly impact the integrated spectral apparent emissivity and this impact is even higher at shorter wavelengths.

Often, the base center temperature of the cavity is chosen as the reference temperature to define the temperature distribution in the literature. In an effort to determine a more realistic reference temperature for the cavity, He et al. [129]. divided the cavity wall into several elements and assigned accurate temperature values to each segment. A weight function for each segment was determined using the fraction of irradiation incident on each segment. The reference temperature was then obtained by summing each segment's temperature multiplied by the weight of each segment.

6.5 Cylindro-inner cone cavity

The literature review in this section only addresses works related to cylindro-inner cone cavities having diffusely reflecting and emitting surfaces. To date, the influence of having specular surfaces on apparent radiative properties has not been explored for this geometry.

A cylindro-inner cone cavity has been shown to have a uniform local apparent emissivity, especially in the conical section. However, due to the complexity of construction and analysis (especially in obtaining the view factor between the conical bottom and a location on the cylindrical body), it has rarely been used as a standard radiation source [130,131]. The combination of a cylinder and cone may result in a higher emission than a cylindro-conical cavity and requires shorter lengths for the cylindrical section. However, one common issue with using cylindro-inner cone cavities is the shadowing effect when a detector is used for the measurement of integrated apparent emissivity. The reentrant cone causes shadowing and results in the invisibility of different locations of the cavity from the detector [31]. When shadowing occurs, the governing integral equations can be simplified using the Stokes theorem, where double integrals are transformed to a single contour integral around the boundary of the cavity surface elements. Then the single integral can be solved through a simple numerical integration.

To enhance emission from a cylindro-inner cone cavity, the addition of a specularly reflecting lid/aperture to a diffusely emitting and reflecting cylindro-inner cone has been evaluated [132]. While effective at increasing the apparent emissivity, having a specularly reflecting lid does not always guarantee a higher apparent emissivity than a lidless cylindro-inner cone cavity.

6.6 Double-cone cavity

Due to the lack of studies considering specularly reflecting double-cone cavities, a summary of past works only for diffusely reflecting and emitting double-cone cavities is provided in this section.

A double-cone cavity has been rarely used as a standard blackbody source or as a standard radiometer. In work by Bedford and Ma [133] it was shown that by increasing the frustum taper the local apparent emissivity monotonically grows and becomes more uniform in the conical section. This behavior is even more significant if the cavity adds a lid/aperture. Changing the conical section angle is another important factor that has been assessed in both lidless and lidded double-cone cavities analyses. For a lidless double-cone cavity, the local apparent emissivity in the conical section and in the frustum experiences a monotonic reduction with increasing conical angle. In the lidded case, however, the change in local apparent emissivity is no longer monotonic but has a deep local minimum near the cavity opening. The emission from a double-cone cavity can be enhanced by adding a cylindrical region to the cavity opening shown in Fig. 10F, also called an extended cone cavity [127].

7. Comparison of hemispherical apparent emissivity

As summarized in Section 6, several different methods under a variety of conditions for obtaining the apparent emissivity of a cavity have been defined and employed in the literature [29,31,43,60,62,68,134]. In this section, we compare the reported apparent emissivity for different cavity geometries considering different surface reflection conditions (i.e., diffuse and specular) and analysis approaches. The tabulated information includes geometric properties, intrinsic emissivity, methods employed, and citations. Results of the closed-form solutions obtained in Section 5 for two-surface enclosures with opaque, isothermal, diffuse, and gray walls are also included in the table for comparison and to indicate the limitations associated with modeling as a two-surface enclosure. It should be noted that the difference observed between the prediction obtained by employing a two-surface enclosure and that obtained by others in the literature is not an indication of error in the modeling approach of this work. Indeed, in Section 4, we showed that subdividing the cavity walls into numerous elements but using the same modeling approach results in precise values for the total hemispherical apparent emissivity. Rather, we provide here a comparison of the two-surface cavities with published values to quantify the disparity associated with such an assumption when occasion warrants the simplicity of using a closed-form expression to predict total hemispherical apparent emissivity. Further, it is important to note that the apparent emissivity has been quantified in the literature for a broader range of geometric and material properties than the tabulated data. However, to ensure a meaningful comparison, only the pertinent published values are presented in this section.

Table 1 summarizes published values for the total hemispherical and total normal apparent emissivity of a cylindrical cavity with opaque, diffusely emitting and gray surfaces over a range of L/D ratios and intrinsic emissivity values. It is observed that the total normal and total hemispherical apparent emissivity exhibit strong agreement at higher intrinsic emissivity values (typically $\varepsilon \geq 0.8$). In other words, irrespective of the surface conditions, they can serve as viable alternatives when dealing with high intrinsic emissivity.

The data reported in all columns of Table 1 except for the last column are based on an additional assumption of diffuse reflection. The last column reports the total normal apparent emissivity of an isothermal cylinder but with specular reflection (and diffuse emission, gray behavior). This condition, reported by Su et al. [134]., allows one to compare the influence of assuming diffuse or specular walls on a cylindrical cavity. It is generally observed that a specular component of reflection within a cylindrical cavity leads to a lower total normal apparent emissivity as compared to a diffuse case, except when the length-to-diameter ratio is at its lowest reported ($L/D = 1$). This observation suggests that as the L/D ratio decreases, the influence of the specularity ratio on the apparent emissivity becomes more pronounced.

Table 1 Summary of reported apparent emissivity values for a cylindrical cavity with opaque, isothermal, and gray surfaces. All the reported works in this table consider diffuse emission and reflection surface conditions, except those with an asterisk indicating diffuse emission and *specular* reflection.

Solution method		Analytical (two-surface enclosure)	Numerical solution of integral equations	Numerical solution of integral equations	Analytical	Analytical	MCRT
Geometric properties	Total hemispherical intrinsic emissivity	Total hemispherical apparent emissivity (current work)	Total hemispherical apparent emissivity (Sparrow et al. [68])	Total hemispherical apparent emissivity (Chandos and Chandos [43])	Total normal apparent emissivity (Campanaro and Ricolfi [60])	Total normal apparent emissivity (Gouffe [62])	Total normal apparent emissivity* (Su et al. [134])
$L/D = 4$	$\varepsilon = 0.9$	0.994	0.9749	0.9750	0.998	–	0.996
	$\varepsilon = 0.7$	0.975	–	0.9154	–	–	–
	$\varepsilon = 0.5$	0.944	0.8367	0.8366	0.989	–	0.954
	$\varepsilon = 0.3$	0.879	–	0.7206	–	–	0.885
$L/D = 2$	$\varepsilon = 0.9$	0.988	0.9746	0.9747	0.994	–	0.994
	$\varepsilon = 0.7$	0.955	–	0.9141	–	–	–
	$\varepsilon = 0.5$	0.900	0.8331	0.8329	0.959	0.927	0.913
	$\varepsilon = 0.3$	0.794	–	0.7095	–	–	0.802

$L/D = 1$	$\varepsilon = 0.9$	0.978	0.9720	0.9721	0.982	–	0.985
	$\varepsilon = 0.7$	0.921	–	0.9038	–	–	–
	$\varepsilon = 0.5$	0.833	0.8084	0.8083	0.776	0.843	0.852
	$\varepsilon = 0.3$	0.682	–	0.6566	–	–	0.700

*Specular reflection.

Table 1 includes data from Chandos et al. [43]. and Sparrow et al. [68]. for the apparent emissivity of an isothermal, diffuse, and gray cylindrical cavity as obtained through radiant flux balance modeling on infinitesimal elements of the surfaces comprising the cavity. Their approach was essentially to model the radiosity leaving a specified location of the cavity (on either the cylindrical sidewall or end disk) as the sum of the emission and reflected incident irradiation from this location. This approach resulted in two coupled integral equations (representing the local apparent emissivity for the sidewall and the base) which required iteration to simultaneously obtain the solution for the total hemispherical apparent emissivity. Their reported values exhibit remarkable agreement (likely within numerical error) as they employed similar approaches to quantity the total hemispherical apparent emissivity. In the subsequent discussion, we compare our predictions with Chandos et al. as they considered a cylindrical cavity with the same emission and reflection characteristics in their analysis, as well as for several additional cavity shapes.

We now provide a comparison between the results of the current approach assuming a two-surface enclosure and that reported by Chandos et al. [43]. The comparison reveals that the maximum disparity between the two works is approximately 22%, primarily occurring at the highest reported L/D ratio of 4 and the lowest intrinsic emissivity of 0.3. Conversely, the smallest difference of 0.63% was observed at the smallest L/D ratio of 1 and the highest intrinsic emissivity of 0.9. Notably, a consistent trend emerges where, regardless of the L/D ratio, increasing the intrinsic emissivity leads to a better agreement between the two-surface enclosure assumption that reported by Chandos et al. [43]. Despite its simplicity, the two-surface enclosure assumption employed in our study yields reasonably accurate results, particularly at higher intrinsic emissivity values, irrespective of the cavity depth.

Additionally, at high intrinsic emissivity values and low L/D ratios, the disparity associated with assuming a two-surface enclosure is diminished. This means, the analytical expressions derived for the apparent emissivity of a two-surface, diffusely emitting and reflecting cavity not only offer convenience in implementation but may also offer acceptable accuracy under certain conditions. Further, one could return to the modeling approach in Section 3 and further subdivide surfaces (as shown in Section 4), to improve accuracy at some cost to efficiency. Notably, it was demonstrated that subdividing and discretizing a cylindrical cavity to account for the actual nature of the radiosity led to an outstanding agreement (with a difference of less than 0.06% between the calculations of this work and reported data by Chandos et al. [43]). Alternatively, the analytical approach of Section 3 can be combined with any numerical or computational technique when additional accuracy is needed.

This table also incorporates the reported total normal apparent emissivities for an isothermal, diffuse, and gray cylindrical cavity by Campanaro et al. [60]. and Gouffe [62]. Campanaro et al.'s approach involved modeling the radiation leaving the cavity base by direct emission and reflection using two disk-shaped elements involved in the successive reflections. They used the concept of an exchange factor (as described in Section 6.1.2) to model the direct emission and reflected irradiation originating from the cavity base and derived a single general formula that applies to both cylindrical and conical cavities. In contrast, Gouffe's methodology (see Section 6.1.1) assumes a uniform distribution of reflected irradiation within the cavity and is a primary limitation of this study. In this work, the fraction $(\rho\Omega/\pi)$ of irradiation reflected after the first reflection in terms of π the opening surface and cavity surface areas was determined. This fraction was subsequently employed to compute the total normal apparent emissivity.

In all tables within this section, a consistent observation emerges regarding the data points pertaining to various cavity geometric shapes: an increase in the intrinsic emissivity and L/D ratio leads to a corresponding increase in both total hemispherical and total normal apparent emissivity. Irrespective of the method employed, the total hemispherical and total normal apparent emissivities reported in the literature align with the analytically derived values in the current study, particularly for high intrinsic emissivity values.

Table 2 summarizes published values for the apparent emissivity of a conical cavity with isothermal, opaque, diffuse, and gray surfaces over a range of L/D ratios and intrinsic emissivity values. The data in this table demonstrates the changes in the apparent emissivity of the conical cavity as L/D and the intrinsic emissivity vary. Similar to the cylindrical cavity, data reported by various publications confirm that, regardless of the nature of emission and reflection, increasing the L/D and intrinsic emissivity increase the apparent emissivity. It is also seen that the total hemispherical and total normal apparent emissivity are in close agreement at higher intrinsic emissivities. Also, specular components in reflections significantly impact the total normal apparent emissivity, yet their presence does not always guarantee a higher total normal apparent emissivity compared to a diffuse case. For instance, in the case of a diffusely reflecting conical cavity, the reported data by Campanaro et al. [60]. indicates a slightly lower total

Table 2 Summary of reported apparent emissivity values for a conical cavity with opaque, isothermal, and gray surfaces. All the reported works in this table consider diffuse emission and reflection surface conditions, except those with an asterisk indicating diffuse emission and *specular* reflection.

Solution method:		Analytical (two-surface enclosure)	Numerical solution of integral equations	Analytical	Analytical	MCRT
Geometric properties	Total hemispherical intrinsic emissivity	Total hemispherical apparent emissivity (current work)	Total hemispherical apparent emissivity (Chandos and Chandos [43])	Total normal apparent emissivity (Campanaro and Ricolfi [60])	Total normal apparent emissivity (Gouffe [62])	Total normal apparent emissivity* (Su et al. [134])
$L/D = 2$	$\varepsilon = 0.9$	0.974	0.9595	0.977	0.991	0.981
	$\varepsilon = 0.7$	0.906	0.8641	0.922	0.957	0.937
	$\varepsilon = 0.5$	0.805	0.7406	0.850	0.886	0.888
	$\varepsilon = 0.3$	0.639	0.5668	0.760	0.741	0.832
$L/D = 1$	$\varepsilon = 0.9$	0.947	0.9421	0.945	0.972	0.960
	$\varepsilon = 0.7$	0.824	0.8107	0.823	0.897	0.873
	$\varepsilon = 0.5$	0.667	0.6512	0.683	0.781	0.775
	$\varepsilon = 0.3$	0.462	0.4498	0.520	0.595	0.677

$L/D = 0.5$	$\varepsilon = 0.9$	0.927	0.9255	0.923	0.948	0.936
	$\varepsilon = 0.7$	0.767	0.7639	0.761	0.827	0.800
	$\varepsilon = 0.5$	0.586	0.5821	0.589	0.679	0.651
	$\varepsilon = 0.3$	0.377	0.3751	0.404	0.478	0.489

*Specular reflection

normal apparent emissivity compared to that reported by Sue et al. [134]., who studied a conical cavity having specular surfaces. In contrast, the reported data from Gouffe [62] generally shows higher values than the data reported by Sue et al. [134]., especially at high intrinsic emissivity values.

The current work, assuming a two-surface enclosure, agrees reasonably well with the data reported by Chandos et al. in a conical cavity, particularly at high intrinsic emissivity values. However, as the intrinsic emissivity decreases, this agreement tends to degrade. Notably, at the highest reported L/D ratio of 4 and the lowest intrinsic emissivity of 0.3, the maximum disparity between our study and Chandos et al.'s work is approximately 13%. On the other hand, the smallest difference between our findings and Chandos et al. work is observed at the lowest presented L/D ratio of 1 and the highest intrinsic emissivity of 0.9, amounting to only 0.18% [43]. Additionally, regardless of the L/D ratio, an increase in the intrinsic emissivity leads to a stronger agreement between the two approaches.

Table 3 summarizes published values for the apparent emissivity of the spherical cavity with opaque, diffusely emitting, and gray surfaces over a range of L/D ratios (i.e., the ratio of the cavity diameter to the opening diameter d/D) and intrinsic emissivity values. Although the total hemispherical and total normal apparent emissivity of spherical cavities has been less extensively studied, it is observed that the presence of specularly reflecting surfaces in such cavities can result in higher total normal apparent emissivity compared to the total hemispherical apparent emissivity, especially when the intrinsic emissivity is less than or equal to 0.5, for any L/D ratio. However, as the intrinsic emissivity increases, this behavior becomes less consistent; for certain higher intrinsic emissivity values, a diffusely reflecting spherical cavity may exhibit a smaller total hemispherical apparent emissivity than a total normal apparent emissivity in a cavity with specular reflections.

Consistent with the findings in cylindrical and conical cavities, in deep spherical cavities the total hemispherical and total normal apparent emissivity agree well. For instance, when comparing the two-surface enclosure assumption from this work for a diffuse spherical cavity with data from Sue et al. [134]. for specular surfaces, the difference is negligible at L/D ratio of 4 and an intrinsic emissivity of 0.9. This similarity can be attributed to the multiple reflections occurring within the cavity due to high L/D ratios, which lead to a diffuse radiation field regardless of the surface conditions. This observation implies that in deep cavities, the placement of the detector to capture the irradiation leaving the cavity may not significantly affect the apparent emissivity measurement.

Table 3 Summary of reported apparent emissivity values for a spherical cavity with opaque, isothermal, and gray surfaces. All the reported works in this table consider diffuse emission and reflection surface conditions, except those with an asterisk indicating diffuse emission and *specular* reflection.

Solution method		Analytical (two-surface enclosure)	Analytical	MCRT
Geometric properties	Total hemispherical intrinsic emissivity	Total hemispherical apparent emissivity (current work)	Total normal apparent emissivity (Gouffe [62])	Total normal apparent emissivity* (Su et al. [134])
$L/D = 4$	$\varepsilon = 0.9$	0.998	–	0.9982
$L/D = 2$		0.993	–	0.9961
$L/D = 1$		0.964	–	0.9898
$L/D = 0.5$		–	–	0.9600
$L/D = 4$	$\varepsilon = 0.75$	0.995	–	0.9931
$L/D = 2$		0.978	–	0.9844
$L/D = 1$		0.900	–	0.9596
$L/D = 0.5$		–	–	0.8820
$L/D = 4$	$\varepsilon = 0.5$	0.984	–	0.9713
$L/D = 2$		0.938	0.944	0.9348
$L/D = 1$		0.750	0.833	0.8459
$L/D = 0.5$		–	–	0.6955

*Specular reflection.

Table 3 further confirms that deep spherical cavities with high intrinsic emissivity tend to exhibit more blackbody-like behavior. Therefore, applying the diffuse approximation for deep spherical cavities is expected to provide reasonably accurate results for the apparent emissivity of spherical cavities when the surface condition is unknown or entirely specular.

Limited works to quantify the apparent emissivity of diffusely emitting and reflecting cylindro-conical and cylindro-inner cone cavities have been conducted. Table 4 contains the available total hemispherical and total normal apparent emissivity associated with the lidless cylindro-conical and cylindro-inner cone cavity considering diffuse emission and reflection. The apparent emissivity, as determined using the equations from Section 5.5 in the current work assuming a two-surface enclosure, is compared with these data for specific geometric conditions and intrinsic emissivity. There is a noticeable agreement between the total hemispherical apparent emissivity obtained in this study and the total normal apparent emissivity reported by Wang et al. [29]. at high intrinsic emissivity values and L/D ratios. However, this agreement diminishes as the intrinsic emissivity decreases.

Given the limited available data concerning cylindro-conical and cylindro-inner cone cavities in the existing literature, it is reasonable to surmise that the observations made with other cavity shapes could potentially apply to these shapes as well. For instance, it may be inferred that the cylindro-conical and cylindro-inner cone cavities might also experience a decline in the agreement between total hemispherical and total normal apparent emissivity when intrinsic emissivity decreases and L/D ratio increases. Regardless, it is clear that high L/D ratios and intrinsic emissivity values exhibit more blackbody-like behavior for these two shapes.

Among the few works in the literature regarding double-cone cavities, none provide the hemispherical apparent emissivity; papers in Section 6.6 provide only local and integrated apparent emissivity values for this shape.

In this section, we have compiled results from the literature, encompassing the total hemispherical and total normal apparent emissivity values of various cavity shapes obtained through different analysis approaches and considering various geometric/material properties. Depending on the specific application and the emission/reflection characteristics involved, it is possible to determine the apparent emissivity of a cavity solely based on its material and geometric properties. Moreover, this tabulated data serves as a valuable resource, allowing for straightforward adjustments of the geometric properties of different cavity shapes to facilitate estimations of the resulting changes in radiative behavior.

Table 4 Summary of reported apparent emissivity values for the isothermal cylindro-inner cone cavity having opaque and gray surfaces. The three reported works in this table consider diffuse emission and reflection surface conditions.

Solution method		Analytical (two-surface enclosure)	Numerical solution of integral equations	MCRT
Geometric properties	Total hemispherical intrinsic emissivity	Total hemispherical apparent emissivity (current work)	Total hemispherical apparent emissivity (Bedford et al. [31])	Total normal apparent emissivity (Wang et al. [29])
Cylindro–conical $L/D = 3$	$\varepsilon = 0.9$	0.992	–	0.9970
	$\varepsilon = 0.8$	0.981	–	0.9934
	$\varepsilon = 0.7$	0.968	–	0.9889
	$\varepsilon = 0.5$	0.929	–	–
Cylindro–inner cone $L/D = 3$	$\varepsilon = 0.9$	0.990	–	0.9975
	$\varepsilon = 0.8$	0.978	–	0.9947
	$\varepsilon = 0.7$	0.962	–	0.9912
	$\varepsilon = 0.5$	0.916	–	–
Cylindro–inner cone $L/D = 4$	$\varepsilon = 0.7$	0.950	0.9955	–

8. Cavity effect applications

Section 8 highlights several applications of radial and angular cavity shapes and identifies the utility of modifying apparent radiative properties in several applications. Traditionally, cavities have been thought to be limited to thermometry, radiometry, and photometry, as well as achieving blackbody radiation. This section offers a possible additional avenue for new applications, including that for dynamic radiative control and thermal management.

Radial cavity geometries have been commonly used as standard blackbody radiators in various applications, such as thermometry, radiometry, and photometry. One primary application for cylindrical cavities is in radiation thermometry, where temperature sensing of erosive materials such as molten steel is desired. Co-axial tubes may be used for this purpose with the temperature of the object approaching 1500 °C in the base, and the temperature difference between the base and the opening of the object is relatively small. The inner layer of the coaxial tubes serves as a blackbody, while the outer layer resists material erosion. Geometric properties of the cavity play critical roles in temperature sensing effectiveness. Usually, the measurement of the local apparent radiative behavior of blackbody cavities in thermometry is favorable [30,135,136]. Conical cavities can also be used for sheet metal temperature sensing. It has been observed that a conical cavity has a better performance in temperature sensing than a truncated sphere or a double-wedge cavity shape, even though a truncated sphere performs better in terms of stray radiation avoidance [106,137].

Another reported application for radial cavity shapes is in concentrated solar power (CSP) systems. To improve the efficiency of CSP plants, various structures for the solar receiver system of power tower plants using the cavity effect have been explored [138,139]. For example, an array of tubes (with either plain or hexagonal apertures) serving as a multi-cavity receiver in a CSP system has exhibited an apparent absorptivity as high as 99.8%, allowing greater utilization of the incident, concentrated solar energy. Fig. 16A illustrates one design where granular media flows over the surface of the cavity and is heated when solar energy is trapped by the hexagonal opening array of the cavities. This arrangement accompanied by optimal geometric properties, can accelerate heat transfer from the cavity, thereby maximizing the solar-to-electric efficiency [45,140]. Cavity designs for increasing the collection of solar heating in dish receiver systems align well with the geometries explored in this review work [141–144]. Fig. 16B illustrates a cylindrical cavity as a receiver in a solar dish collector, coated

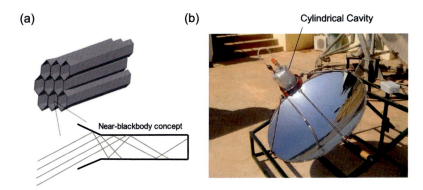

Fig. 16 (A) Hexagonal aperture solar absorber tube with the inset illustrating the cavity effect. (B) A cylindrical cavity receiver in a solar dish collector system [44]. *Adapted from A. Fleming, C. Folsom, H. Ban, Z. Ma, A general method to analyze the thermal performance of multi-cavity concentrating solar power receivers, Sol. Energy. 150 (2015b) 608–618. https://doi.org/10.1016/j.solener.2015.08.007 [45].*

with a highly absorptive black paint [44]. The setup shown in this figure has been designed to efficiently generate process heat above 350 °C for various applications.

Manipulating cavity geometry to achieve favorable radiative performance offers a new avenue of radiative control for applications. The ability to dynamically control the net radiative heat flux through control of radiative properties would enable the thermal management of a component or surface in response to a change in the radiative environment. Intrinsic radiative surface properties are static and, therefore, unable to adapt to changing thermal environments. Thus, for a component operating in conditions where irradiation is time-varying, static intrinsic surface properties may lead to over-heating or excessive heat loss. Grooved cavities of any type (i.e., rectangular, triangular, V-groove, and trapezoidal) have been suggested as efficient structures to enhance absorption and emission in space-bound applications [145–148]. A special type of V-groove cavity with black fins placed at the bottom of a trapezoidal-groove cavity (either with a concentric or linear arrangement of grooves) has shown effective performance in the acceleration of heat transfer and collimation of radiation striking the cavity walls [145,149]. Due to their flexibility in operating in small or large-scale engineering applications such as spacecraft components thermal management and solar steam generators, grooved cavities have recently been a focus of many researchers [3,4,13,14].

Systems or equipment that require thermal management are ubiquitous. One commonly encountered thermal management scenario is indoor climate control. Passive building design strategies often capitalize on solar heating, seasonal solar positioning, and natural ventilation [150–152]. Solar radiation is a primary heating mechanism on the global scale, yet it has a time-varying effect on heating rates (due to diurnal and seasonal solar positions) that is not mitigated or used with static, intrinsic surface properties. Therefore, active indoor climate control strategies, including refrigeration, have become pervasive at a significant cost to energy resources. Recent efforts to achieve daytime radiative cooling have emerged using photonic structures with ordered cavities [153,154]. Further, earth rotation and axial tilt result in solar positioning that requires active tracking for many solar energy systems. 'Kirigami' (cut paper) structures have been proposed to maximize solar utilization through solar tracking [155]. Corrugated panels can also be positioned within the structure to regulate the indoor climate. An innovative corrugated radiant ceiling panel (Fig. 17A) has demonstrated its ability to significantly augment the Energy Efficiency Ratio (EER), resulting in greater cooling capability when compared to conventional flat radiant panels [156].

In addition to these terrestrial applications, 'dynamic' control of thermal systems using radiation is critical in the thermal management of spacecraft and extraterrestrial space stations because radiation is the sole mode of heat exchange with the environment. Dynamically controlling the radiative properties of heat-rejecting systems and minimizing the mass of solid radiator fins in space-based systems has been one of NASA's main challenges [158]. Due to the longevity and commercialization of space flight, the public at large may often conclude that space-based systems are a solved challenge. In reality, the explosion of the internet-of-things and the desire to provide universal access to information is leading to a surge in satellite

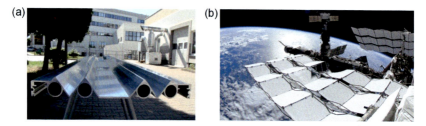

Fig. 17 (A) A new type of monolithic radiant ceiling cooling panel designed with a corrugated surface [156]. (B) Static radiator panels installed on the International Space Station [157].

placement. In addition to the ever-growing demand for national security by remote monitoring, plans for internet-providing satellite networks continue [159].

Satellites, spacecraft, and space exploration have significant associated costs [150–152]. Fig. 17B illustrates radiator panels currently installed on the International Space Station. These panels deploy using rotating fluid joints between panels but remain static during the operation lifetime [157]. Thermal management of spacecraft is a critical aspect of reliable spacecraft design.

Fig. 18 also illustrates two innovative foldable radiant cooling devices with their associated infrared images, created through lamination with integrated microfluidic water-circuits. These devices enhance convective heat transfer due to their unique geometry and increased surface area compared to traditional flat panels. It has been stated that lukewarm water (instead of chilled water) may be sufficient for indoor cooling when employing these devices. This can result in potential savings in energy consumption and the cooling system's lifecycle greenhouse gas emissions [160].

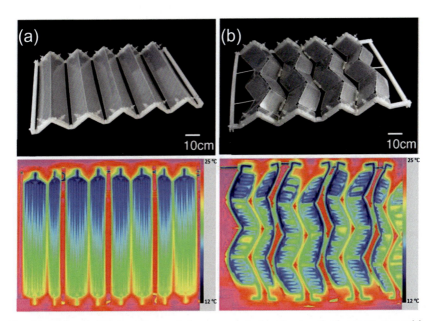

Fig. 18 Radiant ceiling cooling prototypes manufactured using the laminate assembly method to achieve (A) flat panel and (B) zigzag surfaces. *Adapted from J. Grinham, S. Craig, D.E. Ingber, M. Bechthold, Origami microfluidics for radiant cooling with small temperature differences in buildings, Appl. Energy. 277 (2020). https://doi.org/10.1016/j.apenergy.2020.115610.*

Foldable, corrugated surfaces are effective in augmenting emission and reflection. Fig. 19A illustrates an infrared image of a specularly reflecting V-groove cavity, which exhibits the enhanced emission from the cavity in regions where the V-groove angle decreased [9]. An analytical expression for the apparent absorptivity of a V-groove cavity using two methods, namely the integrating factor method (to solve the first order governing transient energy equation for the cavity) and the direct method (obtaining apparent absorptivity using time-dependent temperature measurement), has been derived in the literature [15].

Origami-inspired surfaces, comprised of tessellations, are also able to provide an adaptable surface topography to modify apparent radiative surface behavior. Tessellations are the tiling of a plane with one or more repeated geometric shapes (Fig. 19) and are the building blocks of 3D surfaces that create ordered cavities during the retraction of the folds. Control of the spatial arrangement of tessellations through retraction and expansion offers the ability to reconfigure surfaces and achieve variation in hemispherical and directional radiative surface properties. This change in apparent properties is achieved through the cavity effect. As tessellations that comprise an unfolded surface collapse on each other during folding, grooves are formed, which trap radiative energy due to the high aspect

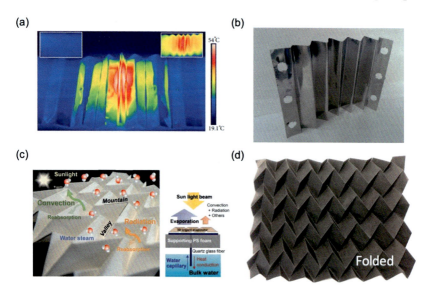

Fig. 19 (A) An infrared image of a folded specularly reflecting V-groove cavity [9], (B) Origami-inspired accordion fold structure [15], (C) Miura-Ori structure used in solar steam generator (the subset demonstrates the mechanism to produce steam via solar steam generator) [3], and (D) folded Baretto Mars structure.

ratio of the cavity. These grooves mimic an isothermal cavity with a small opening, making the surface appear as a 'black' absorber or emitter. Actuating the cavity angle results in control of the degree of black-like behavior, making it possible for a surface to transition between the intrinsic surface property and black behavior [15,161,162].

Cavity geometries also exhibit highly directional characteristics, primarily emitting and absorbing in the direction perpendicular to the projected cavity opening while attenuating emission and absorption in directions near the horizon [105,149,163–168]. Since directionality depends on cavity geometry, origami tessellations also allow for control of directional emission/absorption (ability to transition from diffuse to directional behavior). The 3D nature of origami tessellations gives control of directionality in two dimensions providing a greater degree of collimation.

Origami-inspired surfaces, such as simple accordion folds (Fig. 19B), Miura-ori (Fig. 19C), and Baretto Mars (Fig. 19D), may be used to create a surface topography comprised of cavities with different sizes and shapes. Actuation of an origami-inspired surface to achieve any angle between $0°$ and $180°$ provides the means to transition between the intrinsic surface property and black-like surface behavior. This transition is independent of the intrinsic surface property, with even highly reflective surfaces, achieving black-like behavior at small cavity angles [15,169].

One particular application of Miura-ori (Fig. 19C), has been reported in the solar steam generation process [3]; this system employs a capillary-driven supply of water to the solar panel and transpiration through a water-absorbing glass fiber ribbon. On two rigid parallelogram surfaces, water evaporation takes place. These two surfaces remain intact as the 3D origami is compressed, resulting in an unchanged evaporation rate. The foldability feature of Miura-ori, in turn, enables obtaining high areal density, which is of utmost importance in solar steam generation. The periodic array of mountain-valley folds (in four parallelogram surfaces comprised of Miura-ori where convective and radiative heat exchange may occur) remarkably improves solar energy capture and contributes to recycling the convective and radiative heat loss from irradiated surfaces.

9. Conclusions

In this paper, we explored the cavity effect by studying the influence of geometry on radiative emission from cavities of varying shapes. The following are the main contributions of this work.

- Eq. (6) is a general expression for the spectral, hemispherical apparent emissivity of an arbitrarily shaped cavity comprised of surfaces with spectrally and directionally dependent radiative surface properties. This general expression may be used to find apparent radiative surface properties if the temperature and the spectral, bidirectional reflectivity of each surface comprising the cavity are specified.
- Eq. (19) is an expression of the total, hemispherical apparent emissivity of an arbitrarily shaped cavity. Eq. (19) was verified by calculating the total, hemispherical apparent emissivity of an isothermal, diffusely reflecting and emitting cylindrical cavity, and comparing these results with values available in the literature. Results obtained when sub-dividing the cavity surface into numerous discrete elements exhibited a difference of less than 0.06% when compared to data reported by Chandos et al. [43].
- Eq. (19) was used to develop simple, closed-form expressions for the total, hemispherical apparent emissivity of six representative cavity shapes (i.e., cylindrical, conical, spherical, cylindro-conical, cylindro-inner cone, and double-cone cavities) in which the cavity is modeled as a single surface. These simple two-surface models clearly show the influence of geometric properties (L/D and h/D) and material properties (the intrinsic emissivity) on the apparent emissivity of these cavities. Assuming a two-surface enclosure is a simplification, but it allowed exploration of the effect of geometry on apparent emissivities. We quantified the difference between results based on this assumption and those obtained by other models in the literature to understand its limitations. It was shown that for the same L/D ratio, the spherical cavity exhibits behavior closest to that of a blackbody. Double-cone and cylindro-conical cavities also result in emission close to that of a blackbody. Interestingly, for the same L/D ratio, the conical cavity, which is commonly used to simulate a blackbody, showed the lowest apparent emissivity among these cavity shapes.
- A review of the literature investigating the radiative behavior of these six cavity shapes highlights the influence of geometry on the control of the emissive behavior. Both specular and diffuse conditions were considered in the literature. We also summarize methods employed by other researchers to compute the apparent radiative properties of various cavity shapes, including analytical, computational, and experimental approaches.
- We tabulated the total hemispherical and total normal apparent emissivity values for cavities reported in the literature along with results using

our simplified two-surface analytical model. Further, we compared the reported apparent emissivity values for differing conditions to understand the impact of assuming different surface conditions. It was observed that the presence of specular components in reflections does not always lead to a higher total normal apparent emissivity in a cavity compared to a case with diffuse walls. Interestingly, the impact of specular surfaces on the total normal apparent emissivity becomes more significant at smaller L/D ratios. Moreover, it was observed that the total normal and total hemispherical apparent emissivity were closely aligned for high L/D ratios and intrinsic emissivity values. It was also shown that results obtained using a two-surface enclosure for cylindrical and conical cavities comprised of diffusely emitting and reflecting surfaces differed by less than 22% and 13% from those reported in the literature (for conditions of large L/D and small intrinsic surface emissivity). Factors contributing to the disparities in these results were analyzed, and conditions for which approximating cavities as simple two-surface enclosures are valid are highlighted. In particular, differences in results obtained using simple closed-form expressions and more rigorous methods decreased to less than 0.63% and 0.18% for small L/D ratios or large intrinsic emissivity values.

- We identified applications of the cavity effect in the literature, again highlighting the utility of geometric control of apparent radiative surface properties. Cavities are useful in thermometry, radiometry, and photometry, as well as in simulating blackbody radiation. We extended this view to other topics of interest, such as dynamic control of radiative surface properties that may enable advanced thermal management systems.

Acknowledgments

This review is based upon work supported by the U. S. National Science Foundation (grant number 1749395).

References

[1] A. Prokhorov, N.I. Prokhorova, Application of the three-component bidirectional reflectance distribution function model to Monte Carlo calculation of spectral effective emissivities of nonisothermal blackbody cavities, Appl. Opt. (33) (2012) 8003–8012, https://doi.org/10.1364/AO.51.008003

[2] A.V. Prokhorov, L.M. Hanssen, Effective emissivity of a cylindrical cavity with an inclined bottom: I. Isothermal cavity, Metrologia 41 (2004) 421–431, https://doi.org/10.1088/0026-1394/46/6/C01

[3] S. Hong, Y. Shi, R. Li, C. Zhang, Y. Jin, P. Wang, Nature-inspired, 3D origami solar steam generator toward near full utilization of solar energy, ACS Appl. Mater. Interfaces 10 (2018) 28517–28524, https://doi.org/10.1021/acsami.8b07150

[4] R.B. Mulford, N.S. Collins, M.S. Farnsworth, M.R. Jones, B.D. Iverson, Total hemispherical apparent radiative properties of the infinite V-Groove with diffuse reflection, J. Thermophys. Heat. Trans. 32 (2018) 1108–1112, https://doi.org/10.2514/1.T5485

[5] Z. Yamauti, The light flux distribution of a system of interreflecting surfaces, J. Opt. Soc. Am. 13 (1926) 561–571, https://doi.org/10.1364/JOSA.13.000561

[6] R.B. Mulford, N.S. Collins, M.S. Farnsworth, M.R. Jones, B.D. Iverson, Total hemispherical apparent radiative properties of the infinite V-groove with specular reflection, Int. J. Heat. Mass. Transf. 124 (2018) 168–176, https://doi.org/10.1016/j.ijheatmasstransfer.2018.03.041

[7] W. Biter, S. Hess, S. Oh, D. Douglas, T. Swanson, Electrostatic radiator for satellite temperature control, IEEE Aerosp. Conf. (2005) 781–790, https://doi.org/10.1109/AERO.2005.1559370

[8] K.C. Shannon, H. Demiryont, H. Groger, J. Sheets, A.D. Williams, Thermal management integration using plug-and-play variable emissivity devices, Proc. SPIE 7330, Sens. Syst. Space Appl. III 7330 (2008) 100–108, https://doi.org/10.1117/12.818758

[9] M.J. Blanc, R.B. Mulford, M.R. Jones, B.D. Iverson, Infrared visualization of the cavity effect using origami-inspired surfaces, J. Heat. Transf. 138 (2016) 20901–20902, https://doi.org/10.1115/1.4032229

[10] V. Baturkin, Micro-satellites thermal control - concepts and components, Acta Astronaut 56 (2005) 161–170, https://doi.org/10.1016/j.actaastro.2004.09.003

[11] J.S. Hale, M. Devries, B. Dworak, J.A. Woollam, Visible and infrared optical constants of electrochromic materials for emissivity modulation applications, Thin Solid. Films 313314 (1998), https://doi.org/10.1016/S0040-6090(97)00818-3

[12] S.A. Hill, C. Kostyk, B. Motil, W. Notardonato, S. Rickman, T. Swanson, Draft thermal management systems roadmap, NASA Thermal Management Systems Technology Area Roadmap, 2010.

[13] B.D. Iverson, R.B. Mulford, E.T. Lee, M.R. Jones, Adaptive net radiative heat transfer and thermal management with origami-structured surfaces, Proceedings of the 16th International Heat Transfer Conference, Beijing, China, 2018, pp. 8405–8413.

[14] R.B. Mulford, N.S. Collins, M.S. Farnsworth, M.R. Jones, B.D. Iverson, Total hemispherical apparent radiative properties of the infinite V-groove with diffuse reflection, J. Thermophys. Heat. Trans. 32 (2018) 1108–1112, https://doi.org/10.2514/1.T5485

[15] R.B. Mulford, M.R. Jones, B.D. Iverson, Dynamic control of radiative surface properties with origami-inspired design, J. Heat. Transf. 138 (2016), https://doi.org/10.1115/1.4031749

[16] NASA, View of the aft EEATCS radiator deployment, Archive.Org, 2011. ⟨https://archive.org/details/sts098–335-015⟩ (accessed August 4, 2022).

[17] A. Steinfeld, Solar thermochemical production of hydrogen - a review, Sol. Energy 78 (2005) 603–615, https://doi.org/10.1016/j.solener.2003.12.012

[18] T. Zhao, A. Zhang, G. Mei, S. Zhao, J. Zhang, Effects of misalignment on the accuracy of blackbody cavity sensor for temperature measuring, IEEE Trans. Instrum. Meas. 71 (2022), https://doi.org/10.1109/TIM.2022.3195281

[19] J. Miguel Lopez-Higuera, F.J. Madruga Saavedra, D.A. Gonzalez Fernandez, V. Alvarez Ortego, J. Hierro, High-temperature optical fiber transducer for a smart structure on iron-steel production industry, Proceedings of the SPIE 4328, Smart Structures and Materials 2001: Sensory Phenomena and Measurement Instrumentation for Smart Structures and Materials, 2001. ⟨https://doi.org/10.1117/12.435541⟩.

[20] C.S. Williams, Discussion of the theories of cavity type sources of radiant energy, Appl. Opt. 51 (1961) 564–571, https://doi.org/10.1364/JOSA.51.000564

[21] R.E. Bedford, Effective emissivities of blackbody cavities - a review, Temp. It's Meas. Control. Sci. Ind. 4 (1972) 425–434.

[22] E.R.G. Eckert, E.M. Sparrow, Radiative heat exchange between surfaces with specular reflection, Int. J. Heat. Mass. Transf. 3 (1961) 42–54, https://doi.org/10.1016/0017-9310(61)90004-7

[23] D. Büschgens, C. Schubert, H. Pfeifer, Radiation modelling of arbitrary two-dimensional surfaces using the surface-to-surface approach extended with a blocking algorithm, Heat. Mass. Transf. (2022), https://doi.org/10.1007/s00231-022-03203-4

[24] M.F. Modest, Radiative Heat Transfer, Academic Press, 2013.

[25] T.L. Bergman, A. Lavine, F.P. Incropera, D.P. Dewitt, Fundamentals of Heat and Mass Transfer, John Wiley & Sons, 2011.

[26] J. De Lucas, J.J. Segovia, Uncertainty calculation of the effective emissivity of cylinder-conical blackbody cavities, Metrologia 53 (2015) 61–75, https://doi.org/10.1088/0026-1394/53/1/61

[27] Y. Ohwada, Calculation effective emissivity a cavity having non-Lambertian iso-thermal surface (1999).

[28] E.T. Lee, E. Mofidipour, M.R. Jones, B.D. Iverson, Efficient Monte Carlo ray tracing for apparent cavity behavior, in: Proceedings of the 17th International Heat Transfer Conference, Cape Town, South Africa, August 14–18, 2023.

[29] J. Wang, Z. Yuan, Y. Duan, Comparison of the emissivity uniformity of several blackbody cavities, AIP Conf. Proc. 1552 (8) (2013) 757–761, https://doi.org/10.1063/1.4819637

[30] G. Mei, J. Zhang, S. Zhao, Z. Xie, Simple method for calculating the local effective emissivity of the blackbody cavity as a temperature sensor, Infrared Phys. Technol. 85 (2017) 372–377, https://doi.org/10.1016/j.infrared.2017.07.019

[31] R.E. Bedford, C.K. Ma, Z.X. Chu, Y.X. Sun, S.R. Chen, Emissivities of diffuse cavities.4. Isothermal and non-isothermal cylindro-inner-cones, Appl. Opt. 24 (1985) 2971–2980, https://doi.org/10.1364/AO.24.002971

[32] R.J. Pahl, M.A. Shannon, Analysis of Monte Carlo methods applied to blackbody and lower emissivity cavities, Appl. Opt. 41 (2002) 691–699, https://doi.org/10.1364/AO.41.000691

[33] H. Wann Jensen, L. James Arvo, P. Dutre, A. Keller, A. Owen, M. Pharr, et al., Monte Carlo Ray Tracing, ACM SIGGRAPH. 5 (2003).

[34] A.V. Prokhorov, L.M. Hanssen, S.N. Mekhontsev, Calculation of the radiation characteristics of blackbody radiation sources, Exp. Methods Phys. Sci. 42 (2009) 181–240, https://doi.org/10.1016/S1079-4042(09)04205-2

[35] J.R. Mahan, L.D. Eskin, Applications of Monte Carlo techniques to transient thermal modeling of cavity radiometers having diffuse-specular surfaces, Atmospheric Radiation: 4th Conference, 1981, pp. 181–186. ⟨https://doi.org/10.1207/s15327752jpa8502⟩.

[36] Y.M. Guo, S.J. Pang, Z.J. Luo, Y. Shuai, H.P. Tan, H. Qi, Measurement of directional spectral emissivity at high temperatures, Int. J. Thermophys. 40 (2019), https://doi.org/10.1007/s10765-018-2472-2

[37] G. Keresztury, J. Mink, J. Kristof, Quantitative aspects of FT-IR emission spectro-scopy and simulation of emission-absorption spectra, Anal. Chem. 67 (1995) 1747–1749, https://doi.org/10.1021/ac00116a026

[38] K. Yu, H. Zhang, Y. Liu, Y. Liu, Study of normal spectral emissivity of copper during thermal oxidation at different temperatures and heating times, Int. J. Heat. Mass. Transf. 129 (2019) 1066–1074, https://doi.org/10.1016/j.ijheatmasstransfer.2018.09.116

[39] J. Dai, X. Wang, G. Yuan, Fourier transform spectrometer for spectral emissivity measurement in the temperature range between 60 and 1500°C, J. Phys. Conf. Ser. 13 (2005) 63–66, https://doi.org/10.1088/1742-6596/13/1/015

[40] K.S. Meaker, E. Mofidipour, M.R. Jones, B.D. Iverson, Measured spectral, directional radiative behavior of corrugated surfaces, Int. J. Heat. Mass. Transf. 202 (2023) 123745, https://doi.org/10.1016/j.ijheatmasstransfer.2022.123745

[41] A. Seifter, K. Boboridis, A.W. Obst, Emissivity measurements on metallic surfaces with various degrees of roughness: a comparison of laser polarimetry and integrating sphere reflectometry, Int. J. Thermophys. 25 (2004), https://doi.org/10.1023/B:IJOT.0000028489.81327.b7

[42] K. Gindele, M. Kohi, M. Mast, Spectral reflectance measurements using an integrating sphere in the infrared, Appl. Opt. 24 (1985) 1757–1760, https://doi.org/10.1364/AO.24.001757

[43] R.J. Chandos, R.E. Chandos, Radiometric properties of isothermal diffuse wall cavity sources, Appl. Opt. 13 (1974) 2142–2152.

[44] K. Mahdi, N. Bellel, Development of a spherical solar collector with a cylindrical receiver, in: Energy Procedia, Elsevier Ltd, 2014, pp. 438–448. ⟨https://doi.org/10.1016/j.egypro.2014.07.096⟩.

[45] A. Fleming, C. Folsom, H. Ban, Z. Ma, A general method to analyze the thermal performance of multi-cavity concentrating solar power receivers, Sol. Energy 150 (2015) 608–618, https://doi.org/10.1016/j.solener.2015.08.007

[46] E.M. Sparrow, V.K. Jonsson, Absorption and emission characteristics of diffuse spherical enclosures, J. Heat. Transf. 84 (1962) 188–189.

[47] R.D. Jilte, S.B. Kedare, J.K. Nayak, Natural convection and radiation heat loss from open cavities of different shapes and sizes used with dish concentrator, Mech. Eng. Res. 3 (2013) 25, https://doi.org/10.5539/mer.v3n1p25

[48] Z. Yuan, K. Yu, L. Li, G. Wang, K. Zhang, Y. Liu, New directional spectral emissivity measurement apparatus simultaneously collecting the blackbody and sample radiation, 044902, Rev. Sci. Instrum. 93 (2022), https://doi.org/10.1063/5.0073459

[49] F. Zhang, K. Yu, K. Zhang, Y. Liu, K. Xu, Y. Liu, An emissivity measurement apparatus for near infrared spectrum, Infrared Phys. Technol. 73 (2015) 275–280, https://doi.org/10.1016/j.infrared.2015.10.001

[50] K. Zhang, Y. Xu, X. Wu, K. Yu, Y. Liu, A new approach for accurately measuring the spectral emissivity via modulating the surrounding radiation, J. Quant. Spectrosc. Radiat. Transf. 288 (2022) 108277, https://doi.org/10.1016/j.jqsrt.2022.108277

[51] M. Ballico, A simple technique for measuring the infrared emissivity of black-body radiators, Metrologia 37 (2000) 295, https://doi.org/10.1088/0026-1394/37/4/5

[52] J. Ishii, A. Ono, Uncertainty estimation for emissivity measurements near room temperature with a Fourier transform spectrometer, Meas. Sci. Technol. 12 (2001) 2103–2112, https://doi.org/10.1088/0957-0233/12/12/311

[53] T.J. Quinn, The calculation of the emissivity of cylindrical cavities giving near blackbody radiation, Br. J. Appl. Phys. 18 (1967) 1105–1113, https://doi.org/10.1088/0508-3443/18/8/310

[54] S.R. Chen, Z.X. Chu, H.P. Chen, Precise calculation of the integrated emissivity of baffled blackbody cavities, Metrologia 16 (1980) 69–72.

[55] Z. Chu, R.E. Bedford, W. Xu, X. Liu, General formulation for the integrated effective emissivity of any axisymmetric diffuse blackbody cavity, Appl. Opt. 28 (1989) 1826–1829, https://doi.org/10.1364/AO.28.001826

[56] F.J. Kelly, An equation for the local thermal emissivity at the vertex of a diffuse conical or v groove cavity, Appl. Opt. 5 (1966) 925–927, https://doi.org/10.1364/AO.5.000925

[57] R.E. Bedford, C.K. Ma, Emissivities of diffuse cavities, II: isothermal and non-isothermal cylindro-cones, J. Opt. Soc. Am. 64 (1974) 339–349, https://doi.org/10.1364/JOSA.64.000339

[58] H. Buckley, On the radiation from the inside of a circular cylinder III, Philos. Mag. 17 (1927) 753–762.

[59] M.G. Rossmann, Radiation from a hollow cylinder, Br. J. Appl. Phys. 6 (1955) 262–264, https://doi.org/10.1088/0508-3443/6/7/112

[60] P. Campanaro, T. Ricolfi, New determination of the total normal emissivity of cylindrical and conical cavities, J. Opt. Soc. Am. 57 (1967) 48–50, https://doi.org/10.1364/JOSA.57.000048

[61] F.E. Nicodemus, Directional reflectance and emissivity of an opaque surface, Appl. Opt. 4 (1965) 767–773, https://doi.org/10.1364/AO.9.001474

[62] A. Gouffe, Corrections douverture des corpsnoirs artificiels compte tenu des diffusions multiples internes, Rev. D'Optique 24 (1945) 1–7.

[63] E.W. Treuenfels, Emissivity of isothermal cavities, J. Opt. Soc. Am. 53 (1963) 1162–1171, https://doi.org/10.1364/JOSA.53.001162

[64] B.A. Peavy, A note on the numerical evaluation of thermal radiation characteristics of diffuse cylindrical and conical cavities, J. Res. Natl Bur. Standards-C. Eng. Instrumentation. 70 (1966) 139–147.

[65] E.M. Sparrow, L.U. Albers, Apparent emissivity and heat transfer in a long cylindrical hole, J. Heat. Transf. 82 (1960) 253–255, https://doi.org/10.1115/1.3679925

[66] J. Geist, Note on the quality of freezing point blackbodies, Appl. Opt. 10 (1971) 2188–2190, https://doi.org/10.1364/AO.10.2188_1

[67] J. Geist, Theoretical analysis of laboratory blackbodies. 1: a generalized integral equation, Appl. Opt. 12 (1973) 1325–1330, https://doi.org/10.1364/AO.12.001325

[68] E.M. Sparrow, L.U. Albers, E.R.G. Eckert, Thermal radiation characteristics of cylindrical enclosures, J. Heat. Transf. 84 (1962) 73–81, https://doi.org/10.1115/1.3684295

[69] J. Vollmer, Study of the effective thermal emittance of cylindrical cavities, J. Opt. Soc. Am. 47 (1957) 926–932, https://doi.org/10.1364/JOSA.47.000926

[70] E.M. Sparrow, Radiant emission characteristics of nonisothermal cylindrical cavities, Appl. Opt. 4 (1965) 41–43, https://doi.org/10.1364/AO.4.000041

[71] B. Liu, C.Y. Wang, J.L. Xu, Z.Y. Wang, P.F. Zhu, Research of a blackbody cavity for effective integrated emissivity with finite volume method, IOP Conf. Ser. Mater. Sci. Eng. 721 (2020), https://doi.org/10.1088/1757-899X/721/1/012028

[72] Y. Ohwada, Evaluation of effective emissivities of nonisothermal cavities, Appl. Opt. 22 (1983) 2322–2325, https://doi.org/10.1364/AO.22.002322

[73] C.J. Sydnor, Series representation of the solution of the integral equation for emissivity of cavities, J. Opt. Soc. Am. 59 (1969) 1288–1290, https://doi.org/10.1364/JOSA.59.001288

[74] Y. Ohwada, Numerical calculation of multiple reflections in diffuse cavities, J. Opt. Soc. Am. 71 (1981) 106, https://doi.org/10.1364/JOSA.71.000106

[75] Y. Ohwada, Numerical calculation of bandlimited effective emissivity, Appl. Opt. 24 (1985) 280–283, https://doi.org/10.1364/AO.24.000280

[76] E.M. Sparrow, R.P. Heinisch, The normal emittance of circular cylindrical cavities, Appl. Opt. 9 (1970) 2569–2572, https://doi.org/10.1364/AO.9.002569

[77] W.B. Fussell, Normal emissivity of an isothermal diffusely reflecting cylindrical cavity (with top) as a function of inside radius, J. Res. Natl Bur. Stand. A Phys. Chem. 76A (1972) 347.

[78] Q. Fang, W. Fang, Z. Yang, B.X. Yu, H. Hu, A Monte Carlo method for calculating the angle factor of diffuse cavities, Metrologia 49 (2012) 572–576, https://doi.org/10.1088/0026-1394/49/4/572

[79] A. Ono, Calculation of the directional emissivities of cavities by the Monte Carlo method, J. Opt. Soc. Am. 70 (1980) 547, https://doi.org/10.1364/JOSA.70.000547

[80] Y. Ohwada, H. Sakate, Effective emissivity of an isothermal cylinder having a longitudinal strip opening, Appl. Opt. 26 (1987) 3186–3187, https://doi.org/10.1364/AO.26.003186

[81] E.M. Sparrow, R.P. Heinisch, N. Shamsundar, Apparent hemispherical emittance of baffled cylindrical cavities, J. Heat. Transf. 96 (1974) 112–114, https://doi.org/10.1115/1.3450126

[82] A. Gaetano, Apparent thermal emittance of cylindrical enclosures with and without diaphragms, Int. J. Heat. Mass. Transf. 15 (1972) 2671–2674, https://doi.org/10.1016/0017-9310(72)90156-1

[83] P.D. Kruger, Radiative transfer in cavities with nonisothermal walls, PhD Dissertation., University of Minnesota, 1973.

[84] E.M. Sparrow, Radiation from cavities with nonisothermal heat conducting walls, J. Heat. Transf. 96 (1974) 15–20, https://doi.org/10.1115/1.3450132

[85] C. Hongpan, C. Shouren, C. Zaixiang, The evaluation of the emissivity and the temperature of cavities at the gold freezing point, Metrologia 17 (1981) 59–63, https://doi.org/10.1088/0026-1394/17/2/004

[86] J.C. De Vos, Evaluation of the quality of a blackbody, Physica 20 (1954) 669–689, https://doi.org/10.1016/S0031-8914(54)80181-9

[87] S.H. Lin, E.M. Sparrow, Radiant interchange among curved specularly reflecting surfaces - Application to cylindrical and conical cavities, J. Heat. Transf. 87 (1965) 299, https://doi.org/10.1115/1.3689093

[88] E.M. Sparrow, S.L. Lin, Radiation heat transfer at a surface having both specular and diffuse reflectance components, Int. J. Heat. Mass. Transf. 8 (1965) 769–779, https://doi.org/10.1016/0017-9310(65)90023-2

[89] S.H. Lin, E.M. Sparrow, Absorption characteristics of a specularly reflecting cylindrical cavity irradiated by an obliquely inclined ray bundle, Appl. Opt. 4 (1965) 277–285, https://doi.org/10.1364/AO.4.000277

[90] M. Perlmutter, R. Siegel, Effect of specularly reflecting gray surface on thermal radiation through a tube and from its heated wall, J. Heat. Transf. 85 (1963) 55–62, https://doi.org/10.1115/1.3686010

[91] F.J. Kelly, D.G. Moore, A test of analytical expressions for the thermal emissivity of shallow cylindrical cavities, Appl. Opt. 4 (1965) 31–40, https://doi.org/10.1364/AO.4.000031

[92] J.F. Baumeister, Thermal radiation characteristics of nonisothermal cylindrical enclosures using a numerical ray tracing technique, 5th Thermophysics and Heat Transfer Conference, 1990.

[93] Y. Zhang, J. Dong, Z. Zhang, H. Sun, G. Mei, Circle vector function applied to ray tracing in MCM for calculating effective emissivities of blackbody cavities, Infrared Phys. Technol. 104 (2020), https://doi.org/10.1016/j.infrared.2019.103104

[94] E.M. Sparrow, Radiant emission, absorption, and transmission characteristics of cavities and passages, NASA Spec. Rep. 55 (1965) 103–115.

[95] S.H. Lin, Radiant interchange in cavities and passages with specularly and diffusely reflecting surfaces, PhD Dissertation, University of Minnesota, 1964.

[96] C.S. Williams, Specularly vs diffusely reflecting walls for cavity type sources of radiant energy, J. Opt. Soc. Am. 59 (1969) 249–252, https://doi.org/10.1364/JOSA.59.000249

[97] G. Alfano, A. Sarno, Normal and hemispherical thermal emittances of cylindrical cavities, Am. Soc. Mech. Eng. (Pap.) 97 (1975) 387–390, https://doi.org/10.1115/1.3450384

[98] A.V. Prokhorov, L.M. Hanssen, Effective emissivity of a cylindrical cavity with an inclined bottom: II. Non-isothermal cavity, Metrologia 47 (2010) 33–46, https://doi.org/10.1088/0026-1394/47/1/005

[99] D.L. Dong Liu, Y.D. Yuanyuan Duan, Z.Y. Zhen Yang, Integrated effective emissivity computation for non-isothermal non-axisymmetric cavities, Chin. Opt. Lett. 11 (2013) 022001–022003, https://doi.org/10.3788/col201311.022001

[100] E.M. Sparrow, Radiant absorption characteristics of concave cylindrical surfaces, J. Heat. Transf. 84 (1962) 283–293, https://doi.org/10.1115/1.3684372

[101] Y. Ohwada, A method for calculating the temperature variation along a cavity wall, Meas. Sci. Technol. 2 (1991) 907–911, https://doi.org/10.1088/0957-0233/2/10/003

[102] J.H. Shirley, J.H. Eberly, Local effective emissivity of conical cavities, Appl. Opt. 18 (1979) 3810–3814, https://doi.org/10.1364/AO.18.003810

[103] Y. Ohwada, Numerical calculation of effective emissivities of diffuse cones with a series technique, Appl. Opt. 20 (1981) 3332–3335, https://doi.org/10.1364/AO.20.003332

[104] E.M. Sparrow, V.K. Jonsson, Radiant emission characteristics of diffuse conical cavities, J. Opt. Soc. Am. 53 (1963) 816–821, https://doi.org/10.1364/JOSA.53.000816

[105] L.G. Polgar, J.R. Howell, Directional thermal-radiative properties of conical cavities, Nasa Technical Note D-2904, 1965.

[106] J.C. Krapez, P.G. Cielo, M. Lamontagne, Reflecting-cavity IR temperature sensors: an analysis of spherical, conical, and double-wedge geometries, Infrared Technol. Appl. 1320 (1990) 186, https://doi.org/10.1117/12.22326

[107] J. De Lucas, Validation of a geometrical model for the calculation of the integrated effective emissivity of conical cavities with a lid, Metrologia 52 (2015) 600–612, https://doi.org/10.1088/0026-1394/52/4/600

[108] R.P. Heinisch, E.M. Sparrow, N. Shamsundar, Radiant emission from baffled conical cavities, J. Opt. Soc. Am. 63 (1973) 152–158, https://doi.org/10.1364/JOSA.63.000152

[109] Q. Fang, W. Fang, Y. Wang, X. Ye, B. Yu, New shape of blackbody cavity: conical generatrix with an inclined bottom, Optical Eng. 51 (2012) 086401, https://doi.org/10.1117/1.oe.51.8.086401

[110] M.L. Fecteau, The emissivity of the diffuse spherical cavity, Appl. Opt. 7 (1968) 1363–1364, https://doi.org/10.1364/AO.7.001363

[111] F.E. Nicodemus, Emissivity of isothermal spherical cavity with gray Lambertian walls, Appl. Opt. 7 (1968) 1359–1362, https://doi.org/10.1364/AO.7.001359

[112] M.J. Caola, Radiation from a nonisothermal spherical cavity: an exact solution, Appl. Opt. 40 (2001) 3232–3234, https://doi.org/10.1364/AO.40.003232

[113] A. Steinfeld, Apparent absorptance for diffusely and specularly reflecting spherical cavities, Int. J. Heat. Mass. Transf. 34 (1991) 1895–1897, https://doi.org/10.1016/0017-9310(91)90163-9

[114] P. Campanaro, T. Ricolfi, Effective emissivity of a spherical cavity, Appl. Opt. 5 (1966) 929–932, https://doi.org/10.1364/AO.5.000929

[115] F. Kowsary, J.R. Mahan, Radiative characteristic of spherical cavities with specular reflectivity component, J. Heat. Transf. 128 (2006) 261–268, https://doi.org/10.1115/1.2151196

[116] P. Saunders, Optimising blackbody cavity shape for spatially uniform integrated emissivity, Int. J. Thermophys. 38 (2017), https://doi.org/10.1007/s10765-016-2145-y

[117] J. De Lucas, J.J. Segovia, Uncertainty calculation of the effective emissivity of cylinder-conical blackbody cavities, Metrologia 53 (2016) 61–75, https://doi.org/10.1088/0026-1394/53/1/61

[118] Y. Ohwada, Calculation of the effective emissivity of a cavity having non-Lambertian isothermal surfaces, J. Optical Soc. Am. A 16 (1999) 1059–1065, https://doi.org/10.1364/JOSAA.16.001059

[119] A. Ono, Evaluation of the effective emissivity of reference sources for the radiometrics emissivity measurements, Int. J. Thermophys. 7 (1986) 443–453.

[120] Y. Ohwada, Influence of deviation from Lambertian reflectance on the effective emissivity of a cavity, Metrologia 32 (1996) 713–716, https://doi.org/10.1088/0026-1394/32/6/65

[121] A.V. Prokhorov, V.I. Sapritskii, I.V. Kliger, Statistical modeling of the radiational characteristics of specular-diffuse black-body Models, Teplofizika Vysok. Temperatur 28 (1990) 99–105.

[122] V.I. Sapritsky, A.V. Prokhorov, Calculation of the effective emissivities of specular diffuse cavities by the Monte Carlo method, Metrologia 29 (1992) 9–14, https://doi.org/10.1088/0026-1394/29/1/003

[123] A.S. Nester, J.R. Mahan, Spatial and angular distributions for irradiance from blackbody cavities, Proceedings of SPIE. 4710, 2002. ⟨https://doi.org/10.1117/12.459562⟩.

[124] Y. Zhao, J. Wang, G. Feng, B. Cao, Comparative study on radiation properties of blackbody cavity model based on Monte Carlo method, Int. J. Thermophys. 41 (2020), https://doi.org/10.1007/s10765-020-02648-1

[125] J. Ishii, M. Kobayashi, F. Sakuma, Effective emissivities of black body cavities with grooved cylinders, Metrologia 35 (1998) 175–180, https://doi.org/10.1088/0026-1394/35/3/5

[126] Y. Ohwada, F. Sakuma, L. Ma, Method to determine temperature distribution and intrinsic emissivity of a cavity, J. Opt. Soc. Am. 17 (2000) 1341–1347, https://doi.org/10.1364/JOSAA.17.001341

[127] S. He, C. Dai, Y. Wang, J. Liu, G. Feng, J. Wang, Analysis and improvements of effective emissivities of nonisothermal blackbody cavities, Appl. Opt. 59 (2020) 6977, https://doi.org/10.1364/ao.397229

[128] V.I. Sapritsky, A.V. Prokhorov, Spectral effective emissivities of nonisothermal cavities calculated by the Monte Carlo method, Appl. Opt. 34 (1995) 5645–5652, https://doi.org/10.1364/AO.34.005645

[129] S. He, C. Dai, Y. Wang, J. Liu, Y. Xie, G. Feng, et al., A method for optimizing the reference temperature in the effective emissivity calculation of nonisothermal blackbody cavities, Opt. Express 28 (2020) 29829, https://doi.org/10.1364/oe.404715

[130] K.H. Berry, Emissivity of a cylindrical black-body cavity with a re-entrant cone and face, J. Phys. E 14 (1981) 629–632, https://doi.org/10.1088/0022-3735/14/5/023

[131] Z.X. Chu, S.R. Chen, H.P. Chen, Radiant emission characteristics of isothermal diffuse cylindrical-inner-cone cavities, J. Opt. Soc. Am. 70 (1980) 1270, https://doi.org/10.1364/JOSA.70.001270

[132] Z.X. Chu, Y.X. Sun, R.E. Bedford, C.K. Ma, Precise calculation of effective emissivity of a diffuse cylindro-inner-cone with a specular Lid, Appl. Opt. 25 (1986) 4343–4348, https://doi.org/10.1364/AO.25.004343

[133] R.E. Bedford, C.K. Ma, Emissivities of diffuse cavities.3. Isothermal and non-isothermal double cones, J. Opt. Soc. Am. 66 (1976) 724–730, https://doi.org/10.1364/josa.66.000724

[134] H.B. Su, R.H. Zhang, X.Z. Tang, X.M. Sun, Determination of the effective emissivity for the regular and irregular cavities using Monte-Carlo method, Int. J. Remote. Sens. 21 (2000) 2313–2319, https://doi.org/10.1080/01431160050029594

[135] G. Mei, J. Zhang, X. Wang, S. Zhao, Z. Xie, Spectral and total effective emissivity of a nonisothermal blackbody cavity formed by two coaxial tubes, Appl. Opt. 54 (2015) 3948, https://doi.org/10.1364/ao.54.003948

[136] G. Mei, J. Zhang, S. Zhao, Z. Xie, Effective emissivity of a blackbody cavity formed by two coaxial tubes, Appl. Opt. 53 (2014) 2507–2514, https://doi.org/10.1364/AO.53.002507

[137] H.H. Safwat, Absorption of thermal radiation in a hemispherical cavity, J. Heat. Transf. C92 (1970) 198–201, https://doi.org/10.1115/1.3449632

[138] L. Weinstein, D. Kraemer, K. McEnaney, G. Chen, Optical cavity for improved performance of solar receivers in solar-thermal systems, Sol. Energy 108 (2014) 69–79, https://doi.org/10.1016/j.solener.2014.06.023

[139] C.K. Ho, B.D. Iverson, Review of high-temperature central receiver designs for concentrating solar power, Renew. Sustain. Energy Rev. 29 (2014) 835–846, https://doi.org/10.1016/j.rser.2013.08.099

[140] A. Fleming, Z. Ma, T. Wendelin, H. Ban, C. Folsom, Thermal modeling of a multi-cavity array receiver performance for concentrating solar power generation, Energy Sustainability 56840 (2015), https://doi.org/10.1115/ES2015-49172

[141] J. Yan, Y. Duo Peng, Z. Ran Cheng, Optimization of a discrete dish concentrator for uniform flux distribution on the cavity receiver of solar concentrator system, Renew. Energy 129 (2018) 431–445, https://doi.org/10.1016/j.renene.2018.06.025

[142] C. Zou, Y. Zhang, Q. Falcoz, P. Neveu, C. Zhang, W. Shu, et al., Design and optimization of a high-temperature cavity receiver for a solar energy cascade utilization system, Renew. Energy 103 (2017) 478–489, https://doi.org/10.1016/j.renene.2016.11.044

[143] A. Hassan, C. Quanfang, S. Abbas, W. Lu, L. Youming, An experimental investigation on thermal and optical analysis of cylindrical and conical cavity copper tube receivers design for solar dish concentrator, Renew. Energy 179 (2021) 1849–1864, https://doi.org/10.1016/j.renene.2021.07.145

[144] A. Maurya, A. Kumar, D. Sharma, A comprehensive review on performance assessment of solar cavity receiver with parabolic dish collector, Energy Sour. A Recov. Util. Environ. Eff. 44 (2022) 4808–4845, https://doi.org/10.1080/15567036.2022.2080890

[145] A.V. Prokhorov, L.M. Hanssen, S.N. Mekhontsev, Radiation properties of IR calibrators with V-grooved surfaces, Thermosense XXVIII 6205 (2006) 28–36, https://doi.org/10.1117/12.667557

[146] A.V. Prokhorov, S.N. Mekhontsev, L.M. Hanssen, Emissivity modeling of thermal radiation sources with concentric grooves, High. Temp. High Press. 35–36 (2003) 199–207, https://doi.org/10.1068/htjr093

[147] W.Z. Black, Optimization of directional emission from V-groove and rectangular cavities, J. Heat. Transf. 95 (1973) 31–36, https://doi.org/10.1115/1.3450000

[148] E.M. Sparrow, V.K. Jonsson, Thermal radiation absorption in rectangular groove cavities, J. Appl. Mech. 30 (1963) 237–244, https://doi.org/10.1115/1.3636518

[149] H. Masuda, Directional control of radiation heat transfer by v-groove cavities - collimation of energy in direction normal to opening, J. Heat. Transf. 102 (1980) 563–567, https://doi.org/10.1115/1.3244341

[150] S. Stevanović, Optimization of passive solar design strategies: a review, Renew. Sustain. Energy Rev. 25 (2013) 177–196, https://doi.org/10.1016/j.rser.2013.04.028

[151] J.A. Duffie, W.A. Beckman, Solar Engineering of Thermal Processes, Wiley, 2013.

[152] K. Kruzner, K. Cox, B. Machmer, L. Klotz, Trends in observable passive solar design strategies for existing homes in the U.S, Energy Policy 55 (2013) 82–94, https://doi.org/10.1016/j.enpol.2012.10.071

[153] A.P. Raman, M.A. Anoma, L. Zhu, E. Rephaeli, S. Fan, Passive radiative cooling below ambient air temperature under direct sunlight, Nature 515 (2014) 540–544, https://doi.org/10.1038/nature13883

[154] E. Rephaeli, A. Raman, S. Fan, Ultrabroadband photonic structures to achieve high-performance daytime radiative cooling, Nano Lett. 13 (2013) 1457–1461, https://doi.org/10.1021/nl4004283

[155] A. Lamoureux, K. Lee, M. Shlian, S.R. Forrest, M. Shtein, Dynamic kirigami structures for integrated solar tracking, Nat. Commun. 6 (2015), https://doi.org/10.1038/ncomms9092

[156] J. Wojtkowiak, Ł. Amanowicz, T. Mróz, A new type of cooling ceiling panel with corrugated surface—Experimental investigation, Int. J. Energy Res. 43 (2019) 7275–7286, https://doi.org/10.1002/er.4753

[157] NASA, International Space Station, n.d. http://spaceflight.nasa.gov/gallery/images/shuttle/sts-135/hires/iss028e016140.jpg (accessed May 5, 2023).

[158] B.N. Tomboulian, R.W. Hyers, Predicting the effective emissivity of an array of aligned carbon fibers using the reverse monte carlo ray-tracing method, J. Heat. Transf. 139 (2017), https://doi.org/10.1115/1.4034310

[159] R. McCormick, The Verge, 2017, SpaceX plans to launch first internet-providing satellites in 2019, 2019. 〈https://www.theverge.com/2017/5/4/15539934/spacex-satellite-internet-launch-2019〉 (accessed May 5, 2023).

[160] J. Grinham, S. Craig, D.E. Ingber, M. Bechthold, Origami microfluidics for radiant cooling with small temperature differences in buildings, Appl. Energy 277 (2020), https://doi.org/10.1016/j.apenergy.2020.115610

[161] X. Wang, H. Jiang, Design of origami fin for heat dissipation enhancement, Appl. Therm. Eng. 145 (2018) 674–684, https://doi.org/10.1016/j.applthermaleng.2018.09.079

[162] R.B. Mulford, V.H. Dwivedi, M.R. Jones, B.D. Iverson, Control of net radiative heat transfer with a variable-emissivity accordion tessellation, J. Heat. Transf. 141 (2019), https://doi.org/10.1115/1.4042442

[163] M. Perlmutter, J.R. Howell, A strongly directional emitting and absorbing surface, J. Heat. Transf. 85 (1963) 282–283, https://doi.org/10.1115/1.3686103

[164] W.Z. Black, R.J. Schoenhals, A study of directional radiation properties of specially prepared V-groove cavities, J. Heat. Transf. 9 (1968) 420–428, https://doi.org/10.1115/1.3597537

[165] R.B. Zipin, The directional spectral reflectance of well characterized symmetric v grooved surfaces, Purdue University, 1965.

[166] R.B. Zipin, A preliminary investigation of the bidirectional spectral reflectance of V-grooved surfaces, Appl. Opt. 5 (1966) 1954–1957, https://doi.org/10.1364/AO.5.001954

[167] O.W. Clausen, J.T. Neu, The use of directionally dependent radiation properties for spacecraft thermal control, Astronautica Acta 11 (1965) 328–339, https://doi.org/10.2514/6.1965-430

[168] K.G.T. Hollands, Directional selectivity, emittance, and absorptance properties of vee corrugated specular surfaces, Sol. Energy 7 (1963), https://doi.org/10.1016/0038-092X(63)90036-7

[169] E.M. Sparrow, R.D. Cess, Radiation Heat Transfer, McGraw-Hill, New York, 1978.

CHAPTER TWO

Hyperthermia applications in cardiovascular and cancer therapy treatments

Sanaz Imanlou and Kambiz Vafai[*]

Mechanical Engineering Department, University of California, Riverside, CA, United States
*Corresponding author. e-mail address: vafai@engr.ucr.edu

Contents

1. Introduction	72
2. Hyperthermia in cardiovascular diseases	72
2.1 LDL transport in arterial walls: two approaches	73
2.2 Governing equations	74
2.3 Effects of hypertension and hyperthermia on LDL	77
2.4 Hyperthermia and soret effects on LDL concentration	78
3. Using hyperthermia in cancer treatment applications	79
3.1 Nanoparticle usage for hyperthermia applications	81
3.2 Gold-based nanostructures	81
3.3 Magnetic nanoparticles	84
3.4 Iron oxide nanoparticles	85
3.5 Carbon nanotubes	85
3.6 Heat generation	86
3.7 Specific absorption rate	88
3.8 Governing equations	88
3.9 The impact of blood vessels on temperature patterns	91
4. Conclusions	91
References	92

Abstract

This work provides an overview of the use of hyperthermia in cardiovascular disease and cancer treatments. In the realm of cardiovascular disease, hyperthermia is used for deliberate exposure of blood vessels or the outer body surface to heat, which can have a significant impact on the cardiovascular disease. The Ludwig-Soret effect, wherein low-density lipoprotein (LDL) particles migrate from warmer to cooler regions, plays a pivotal role in hyperthermia. Understanding the heating effect on LDL accumulation between the endothelium and intima unveils insights into atherosclerotic plaque development. External heating accentuates LDL accumulation, while internal heating reduces LDL concentration, leading to possible therapeutic avenues. In the realm of cancer treatment, tumor cells often exhibit higher susceptibility to heat compared to normal cells. Elevating

Advances in Heat Transfer, Volume 57
ISSN 0065-2717, https://doi.org/10.1016/bs.aiht.2024.02.002
Copyright © 2024 Elsevier Inc. All rights are reserved, including those for text and data

71

temperatures in and around tumors can improve the disruption of cancer cells. Employing various nanoparticles like iron oxide, gold-based, and carbon-based, exhibits potential for cancer therapy using hyperthermia method.

1. Introduction

Approximately two and a half millennia ago, the Greek physician Parmenides expressed the belief that with the ability to induce fevers, he could effectively treat all diseases [1–3]. Hyperthermia, stemming from Greek roots "hyper," meaning increased, and "therme," meaning heat, elevates the specific tissues temperature for targeted therapeutic effects [4].

The primary objective of this chapter is to provide a comprehensive understanding of the usage of hyperthermia in cardiovascular and cancer applications. This chapter is divided into two sections. The first section focuses on the corresponding causes of stenosis formation and low-density lipoprotein (LDL) transport in arterial walls, exploring the effect of hyperthermia on LDL concentrations and presenting the governing equations. The second section introduces the application of hyperthermia in cancer therapy, discussing the materials used in this area, heat generation, relevant governing equations, and the impact of blood vessels on the temperature patterns.

2. Hyperthermia in cardiovascular diseases

Cardiovascular diseases (CVD) are the most prevalent non-communicable diseases and rank among the most common causes of global mortality and illness [5–7]. Significant focus has been placed on CVD because of their substantial impact on public health, which is the primary reason for mortality and impairment in the United States and majority of nations globally leading to high healthcare costs [8–12]. Atherosclerotic heart disease is the most prevalent among them [13–15]. Atherosclerotic heart disease arises from the narrowing of an artery segment, known as stenosis, leading to reduced blood flow. This stenosis is primarily caused by the thickening of the tunica intima, a layer within the artery. The accumulation of lipids, proteins, and fibrous connective tissue in the intima results in the formation of atherosclerotic plaques. Although the precise cause of plaque formation remains a subject of ongoing debate, it is well-established that the oxidation of LDL by free radicals in the intima plays a significant role in promoting plaque growth [8,9,13,16–18]. Stangeby and

Ethier [19] research shows that when there's a stenosis in the artery, it leads to a rise in LDL concentration downstream from the stenosis.

The prevailing theory suggests that monocyte white blood cells infiltrate the tunica intima and transform into macrophages. These macrophages absorb oxidized LDL. High-density lipoprotein (HDL) plays a crucial role in removing LDL from macrophages [8]. When HDL is unable to clear all the LDL, foam cells are generated. Additionally, the release of certain chemotactic factors encourages the migration of smooth muscle cells (SMC) from the tunica media to the intima. These SMCs engulf oxidized LDL, further contributing to the formation of foam cells in the tunica intima. This entire process results in the thickening of the tunica intima, giving rise to the formation of a plaque [8,16,20,21].

Atherosclerotic plaques can be categorized as stable or unstable. Stable plaques feature a thick and robust fibrous cap that separates the plaque from the arterial lumen. In contrast, unstable plaques are particularly dangerous because they possess a thin and fragile fibrous cap, making them susceptible to rupture and potentially leading to the formation of blood clots [16].

Investigation on humans, pigeons, and rabbits suggests that the movement of LDL from the bloodstream into the arterial wall relies on both LDL levels in the blood and the permeability of LDL at the boundary between the bloodstream and the artery wall [22]. In recent years, it has been proven vital to slow atherosclerosis to prevent heart issues, with a focus on lowering LDL as a key strategy for those with the disease [23].

2.1 LDL transport in arterial walls: two approaches

There are two well-established approaches for studying the movement of LDL within artery walls. The first approach involves experimental research with animals. In the first method, Calara and colleagues [24] conducted a study using a rat model where they injected unaltered human LDL. This caused the buildup of oxidized LDL in the walls of the arteries. Within 6 hours of administering the unaltered human LDL, they observed the presence of apolipoprotein B and specific markers found on oxidized LDL in the arterial walls. This indicates that the model effectively mimicked the oxidation of LDL within a living organism.

The second approach entails solving transport equations within simulated geometric models. Yang and Vafai [25], Ai and Vafai [26], and Prosi et al. [27] have categorized these models into three types [16]. The first category includes "wall-free models," which simplify the arterial wall and replace it with boundary conditions [28]. Nematollahi and colleagues [29] introduced a mathematical model that doesn't consider the wall and accounts for non-Newtonian behavior. They explored various factors influencing the concentration on the surface within the lumen of a vessel. However, this model is considered oversimplified as it does not account for the crucial concentration profile along the wall [16,28,30].

The second category involves "fluid-wall models," where the arterial wall is represented as a uniform layer [16,30]. Sun et al. [31] employed this model to describe the transport of oxygen and LDL through this layer.

Previously, researchers primarily relied on wall-free and fluid-wall models, but the results were less accurate compared to experimental data. Vafai et al. [16,25,28,30,32–37] introduced a more comprehensive approach involving four layers, the third category. This complex model takes into consideration the various components of the arterial wall, typically including four layers: the endothelium, intima, internal elastic lamina (IEL), and media. This method meticulously examines the heterogeneity of the arterial wall, providing a detailed analysis of each layer, Fig. 1 [16,30,32].

Researchers have explored various shapes of stenosis, but the most widely employed and realistic geometries are the axisymmetric bell-shaped stenosis and the axisymmetric cosine-shaped stenosis such as the generalized axisymmetric stenosis shape, as introduced by Ai and Vafai [26], and Chung and Vafai [35].

2.2 Governing equations

Assuming that the fluid behaves as a Newtonian fluid and that there are no significant volumetric forces at play, it is possible to formulate comprehensive equations for mass, momentum, species, solid-phase momentum, and energy as [39].

$$\nabla \cdot \mathbf{u} = 0 \tag{1}$$

$$\frac{\rho}{\varepsilon}\left(\frac{\partial \mathbf{u}}{\partial t} + \frac{\mathbf{u}}{\varepsilon}\cdot\nabla\mathbf{u}\right) = -\nabla p + \frac{\mu}{\varepsilon}\nabla^2\mathbf{u} - \frac{\mu}{K}\mathbf{u} + \sigma_0 R T \nabla c \tag{2}$$

$$\frac{\partial c}{\partial t} + (1 - \sigma_s)\mathbf{u}\cdot\nabla c = D_{eff}\nabla^2 c - kc + \frac{k_t \rho_f}{TM_f}\nabla\cdot(D_{eff}\nabla T) \tag{3}$$

$$\rho\frac{\partial^2 x}{\partial t^2} = \nabla\sigma_m + f_m \tag{4}$$

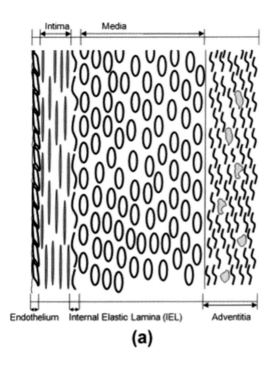

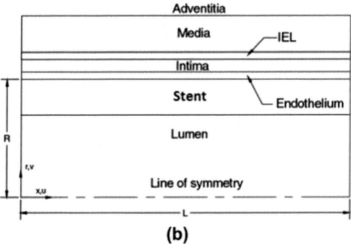

Fig. 1 Schematic of different arterial layers (A) physical display (B) simplified presentation along with the coordinate system [38].

$$\frac{\partial T}{\partial t} + \mathbf{u} \cdot \nabla T = \alpha \nabla^2 T + \frac{RTk_T}{c_T M_f c} \nabla \cdot (D_{eff} \nabla c) \tag{5}$$

These equations apply to values of 'r' within the range from R to $R + R_w$, where 'R' represents the radius of the lumen and 'R_w' denotes the coordinate for arterial wall thickness. As can be seen here, hyperthermia can influence the transport of LDL through various mechanisms, including Ludwig-Soret, thermal expansion, and Dufour effects'. The variables in these equations were volume averaged and denoted by $<>$; however, for simplicity, they are omitted. As seen in Eq. (3), a balance exists between a large-scale inertial term on the left side, while pressure, Brinkman, Darcian, and osmosis effects are considered on the right side. In this equation, K stands for hydraulic permeability, σ_0 represents the osmosis coefficient, while R and T denote the universal gas constant and temperature, respectively. In this context, k_T represents the thermodiffusion coefficient, while M_f stands for molecular mass weight. The term $(1 - \sigma_s)$ signifies the Staverman reflection coefficient, illustrating the solute's rebound at the arterial wall. 'x' denotes displacement, while σ_m and f_m refer to the Cauchy stress tensor and body forces, respectively. α represents the Womersley parameter. The dimensionless velocity 'u' and time 't' are pertinent variables. Additionally, c_T denotes the average LDL concentration [39].

When the outer surface of the body or a blood vessel is exposed to the heat, it causes the inner lining, known as the intima, to become warmer. As a consequence of the Ludwig-Soret effect, LDL particles have a tendency to migrate from regions with higher temperature to those with a lower temperature. This leads to the gathering of LDL particles at the boundary between the endothelium and intima, as the intima maintains a higher overall temperature due to external heating. This accumulation may play a role in the progression of atherosclerosis. In the research carried out by Iasiello M. Vafai K. et al. [40] they provided evidence that when external heating is applied, LDL gathers at the boundary between the endothelium and intima, because of the Ludwig-Soret effect, where particles migrate from warmer to cooler areas, as shown in Fig. 2 Additionally, their findings indicated that the intima exhibits a higher absolute temperature during external heating, which leads to a greater concentration of LDL compared to when internal heating is involved [13,16,35,39–41].

In instances of internal heating, where heat is produced within the body, the intima's temperature might be comparatively cooler than what is observed in external heating situations. This lower intima temperature during internal heating leads to a decreased presence of LDL particles when

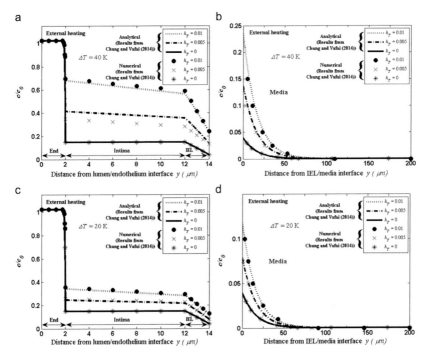

Fig. 2 Concentration profiles for different ΔT's under external heating: (A) and (B) parts are for $\Delta T = 40$ K, while (C) and (D) parts are for $\Delta T = 20$ K [40].

contrasted with external heating. The Ludwig-Soret effect is responsible for diminishing the movement of LDL particles, also known as mass flux, because of the temperature difference [13,16,35,40,41].

In situations involving internal heating, as shown in Fig. 3 there is a tendency for LDL to amass at the interface between the intima and IEL. The intima experiences cooler temperatures during internal heating, resulting in a reduced concentration of LDL as opposed to external heating. The Ludwig-Soret effect, as demonstrated in the study conducted by Iasiello M, Vafai K. and their colleagues, influences mass flux by diminishing it through the impact of temperature [13,16,35,40,41].

2.3 Effects of hypertension and hyperthermia on LDL

The results from Vafai et al. [40] indicate that hypertension typically leads to higher levels of LDL cholesterol in various layers when the tissue is externally heated as shown in Figs. 4 and 5 This increase is further intensified when hyperthermia is involved. When the tissue is internally heated, a similar trend

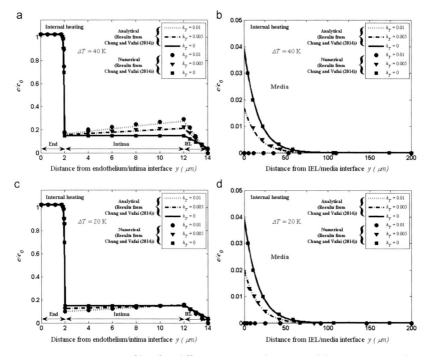

Fig. 3 Concentration profiles for different ΔT's under internal heating: (A) and (B) parts are for $\Delta T = 40$ K, while (C) and (D) parts are for $\Delta T = 20$ K [40].

is observed, with LDL concentration generally rising in each layer among individuals with hypertension. In the endothelium layer, the curves exhibit similar slopes, while in the intima layer, hyperthermia initially reduces LDL concentration in the first part but consistently raises it near the intima/IEL interface compared to the case with no hyperthermia. This reduction followed by an increase is more pronounced in individuals with hypertension. Additionally, the reduction of LDL in the IEL layer appears to be greater when the Ludwig-Soret effect is more noticeable [40].

2.4 Hyperthermia and soret effects on LDL concentration

Based on Chung and studies [13], when external hyperthermia is applied, Soret diffusion boosts the transportation of LDL by increasing its overall concentration in the arterial wall. Conversely, internal hyperthermia may sometimes lead to a decrease in LDL concentration. They noted that the Soret effect within the endothelium layer is less significant because convection and diffusion flux play a more dominant role [13].

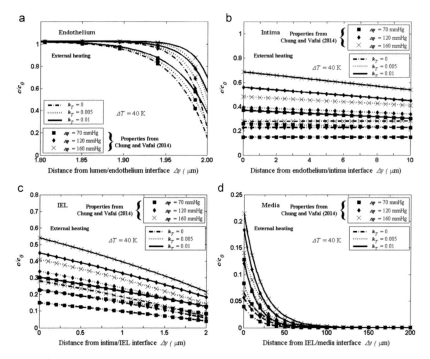

Fig. 4 Concentration profiles in different layers: (A) endothelium, (B) intima, (C) IEL and (D) media under external hyperthermia load, with hypertension effects, for = 40 K [40].

3. Using hyperthermia in cancer treatment applications

Cancer represents a major global threat to the public health and the mortality due to cancer is increasing rapidly recently [42–45] Due to its substantial impact on public health, any progress in cancer treatment can significantly enhance the well-being of individuals affected by the disease. The most commonly employed cancer treatments include chemotherapy, radiation therapy, and surgical intervention [46–49]. Surgical removal of tumors is a highly effective method for eliminating cancerous growths, but it may not always be feasible. On the other hand, while chemotherapy and radiation therapy are effective treatments, their administration at high doses can lead to severe side effects [46–48].

Both experimental and clinical studies have suggested that pairing cytotoxic medications with targeted hyperthermia enhances the destruction of cancerous cells [50]. Hyperthermia alongside chemotherapy enhances

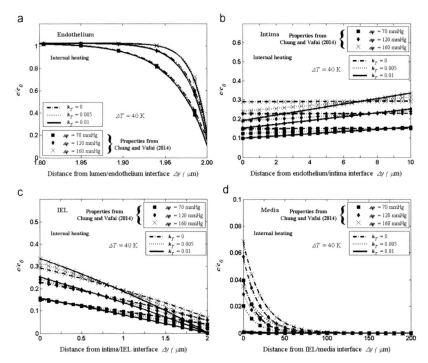

Fig. 5 Concentration profiles in different layers: (A) endothelium, (B) intima, (C) IEL and (D) media under internal hyperthermia load, with hypertension effects, for = 40 K [40].

drug penetration into tumors and boosts the drug's effectiveness against various diseases, including breast cancer, cervical and bladder cancer, rectal cancer, prostate cancer, head and neck cancer, superficial tumors, lung and stomach cancer, and pancreas and liver metastases [51].

Hyperthermia represents an innovative and minimally invasive approach to cancer treatment with limited adverse effects. It involves raising the temperature of tumor tissues, typically to the range of 41–46 °C, leading to the induction of apoptosis or necrosis in cancer cells [47,51–54]. This process is designed to eradicate malignant cells while minimizing damage to nearby healthy tissue caused by overheating [46,47,52,55].

The use of nanoparticles (NPs) for hyperthermia is divided into two groups: magnetic hyperthermia (MHT) and photothermal therapy (PTT). MHT utilizes magnetic NPs within an alternating magnetic field (AMF), while PTT employs light-responsive NPs triggered by a near-infrared (NIR) laser [56].

The irreversible damage to cells occurs because high temperatures lead to the denaturation of proteins. Among the various methods of hyperthermia, MHT treatment employs magnetic nanoparticles (MNPs) injected into the tumor site. These MNPs are then exposed to an AMF to generate the required heat in specific target areas. This mechanism of heat generation offers several advantages over conventional methods, including more precise and uniform heating, safety for the human body, and improved treatment effectiveness [46,47,52].

Targeted hyperthermia involves the utilization of nanoscale metallic particles that can convert electromagnetic energy into heat. This approach offers significant prospects and hurdles for treating cancer and other temperature-sensitive diseases without invasive procedures. The key lies in ensuring that these NPs are taken up by cancerous cells while sparing healthy ones, enabling the precise delivery of electromagnetic energy. The efficiency of electromagnetic energy conversion into heat depends on several physical factors unique to the specific metal in use. The conversion of electromagnetic energy into heat is contingent on various physical factors specific to the type of metal NPs and their electrodynamic response [57].

3.1 Nanoparticle usage for hyperthermia applications

Engineered NPs have emerged as pivotal elements in nanotechnology recently [58]. NPs are typically characterized as solid particles with sizes ranging from 10 to 1000 nanometers, although the European Commission's definition broadens this to encompass materials in which at least half of the particles are 100 nanometers or smaller [59]. Furthermore, NPs frequently display unique and discernable electrical, optical, magnetic, biological, and chemical properties [59,60].

More recently, advancements in hyperthermia treatment have been made through the utilization of nanoscale technologies [61]. Table 1 shows the use of nanomaterials in various applications [59,62]. Particularly, gold-based nanoparticles (AuNPs), carbon-based nanoparticles (CNPs), and iron oxide nanoparticles (IONPs) have emerged as promising nano-sized constructs for enhancing hyperthermia therapy. AuNPs and CNPs can absorb NIR light, making them suitable for tumor PTT [61].

3.2 Gold-based nanostructures

Gold-based nanostructures have been a focus in nano biomedicine since 1950s due to their low toxicity, biocompatibility, adjustable surface plasmon

Table 1 Diverse applications enabled by nanoparticles [59,62].

Major material	Drugs, functionalizing agents, or bioactive molecules	Role and consequent	References
Au	Hesperidin	Antioxidant agent; protective agent against DNA damage from H_2O_2, anticancer drug-delivery system for breast cancer therapy	[118]
Au@Pt	Methoxy-PEG-thiol	PTT under 808 nm light irradiation	[119]
Au	MTX	High therapeutic efficacy; low dose-dependent adverse effect; targeting ability; PTT; PDT; controlled release	[120,121]
Polysulfated gold	–	Optoacoustic tomography; high inflammation targeting potential; imaging inflammation	[122]
AgCu	Sodium citrate, mercapto- propionic acid	anticancer agent	[123]
Pt	Doxorubicin, fucoidan	PTT; Drug delivery	[124]
Pd	Polyvinylpyrrolidone	Anticancer agent, ROS generation and cleaving of the mitochondrial membrane	[125]
Porphyrins	–	PDT; selective accumulation in the inflamed synovial tissues	[126]

Benzoporphyrin derivative monoacid ring- A (BPD-MA)	–	PDT; selective with no harm to surrounding normal tissues	[127]
Tetera suplhonatophenyl porphyrin (TSPP)	–	PDT; effectively accumulate on RA synovial fibroblasts	[128]
ZnO	Quercetin	Drug-delivery system	[129]
Fe_3O_4	BSA protein, glutaric acid	Magnetic hyperthermia agent	[130]
$Fe_3O_4@SiO_2$	lactoferrin, doxorubicin	Drug-delivery system, magnetic hyperthermia, and photothermal agent	[131]
AlMg layered double hydroxide- Fe_3O_4	Hyaluronic acid, doxorubicin	Drug-delivery system; MRI contrast	[132]
Superparamagnetic iron oxide	FA	Negative synovium enhancement and hyperintense; visualizing the interactions with RA state	[133]
Indocyanine green (ICG)	–	PDT; Sonodynamic therapy; cytotoxic effects; apoptosis	[134]
Au@SiO2	Poly (N- isopropylacrylamide- co- acrylic acid); indocyanine green	PTD; PTT; Drug-delivery system	[135]
Silica-carbon nano-onion	Fucoidan, doxorubicin, HM30181A	PTT; Drug-delivery system	[136]
Ag@SiO Ag@SiO@Ag	Not Applicable	PTT; bioimaging agent	[137]

resonance, resistance to oxidation, ease of surface modification, and strong plasmonic photothermal properties [48].

Fig. 6 shows the effects of RF-induced hyperthermia on Panc-1 cells treated with different types of AuNPs. RF treatment alone had low cytotoxicity (gray bar fx1). Cells treated with unlabeled AuNPs (red bars fx2) or IgG-conjugated AuNPs (green bars fx3) did not significantly increase RF-induced thermal cytotoxicity. However, when Panc-1 cells were treated with C225-conjugated AuNPs (dark blue bars fx4), there was a substantial enhancement in RF-induced thermal cytotoxicity, which was linked to increased intracytoplasmic vesicles. This enhanced cytotoxicity with C225-conjugated AuNPs was blocked when Panc-1 cells were first incubated with C225 alone (not conjugated to AuNPs) 30 min before adding the C225-conjugated AuNPs (light blue bar fx5) [57].

3.3 Magnetic nanoparticles

MNPs have garnered significant attention in recent years because of their distinctive physicochemical characteristics and promising biomedical uses, including targeted drug delivery [63–65], magnetic resonance imaging (MRI) [66–69], diagnosis [70], bio-separation [71,72], in brain capillaries for understanding neurodegenerative disorders [73], MHT [74–76], and magnetic storage media [77].

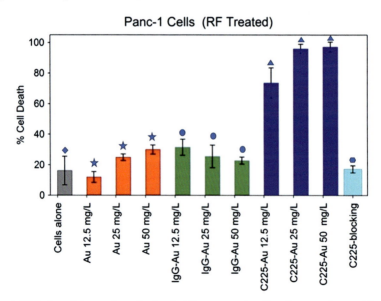

Fig. 6 RF-induced hyperthermia in AuNPs-targeted cells after 2-minute RF field treatment [57].

3.4 Iron oxide nanoparticles

Based on numerous investigations, IONPs, especially superparamagnetic iron oxide nanoparticles (SPIONs) stand as the ones officially approved by the FDA (US Food and Drug Administration) [78,79]. SIONPs can generate heat when subjected to an AMF. IONPs offers two distinct advantages in comparison with CNPs and AuNPs. IONPs allow for virtually unlimited penetration depth within tumor tissue, and they can be readily detected within the human body using clinical MRI [61]. The level of heat generated by MNPs is determined by Rosensweig's model, which is closely associated with the intensity and frequency of AMF [52,80].

Magnetite (Fe_3O_4) is widely favored as a candidate for MNPs in numerous research studies due to its advantageous magnetic characteristics and minimal toxicity [81–83]. Also, manganese ferrite and zinc ferrite are explored due to their strong magnetic properties and their ability to resist oxidation [84].

The hysteresis losses in magnetic particle hyperthermia are noticeably influenced by the size distribution of MNPs [85]. According to Li et al. [86], MNPs ranging from 8 to 24 nm in size are considered effective for treating cancer through in situ hyperthermia when exposed to an alternating current (AC) magnetic field. Among these, MNPs with a 24 nm size exhibited the highest heating efficiency in the applied magnetic field (300 oersted) at a 100 kHz frequency [86]. Atsumi et al. [87]examined magnetic materials for hyperthermia, focusing on their magnetic properties, treatment limitations, and activation methods. They determined that superparamagnetic magnetite with a diameter of 11–13 nm was the most promising choice due to its magnetic and biocompatible attributes [87].

3.5 Carbon nanotubes

Due to the remarkable properties of carbon naotubes (CNTs), such as their low toxicity, stability, and exceptional thermal conductivity, they significantly prolong heat retention within tumor cells, making them essential in hyperthermia treatment [88]. CNTs have been employed for the delivery of genes or proteins into cancer cells through a non-targeted endocytosis process [57]. Studies suggest that when CDs are paired with other nanomaterials, they can amplify PTT efficiency [89]. Behnam et al. [90] research shows that combining CNTs with silver nanoparticles in plasmonic PTT effectively destroyed melanoma tumors. The merging of CNTs with silver bolstered the optical absorption of CNTs, thereby amplifying their efficacy in PPTT for cancer treatment.

Despite its demonstrated effectiveness, MHT treatment still faces a significant challenge: how to cause lethal thermal damage to cancerous tissues while minimizing harm to the surrounding healthy tissue. Overcoming this challenge requires determining the optimal concentrations and spatial distributions of MNPs, which, in turn, necessitates a comprehensive analysis of heat and mass transfer processes [46,47].

3.6 Heat generation

Heating occurs within magnetic particles when they are exposed to a strong magnetic field, which can occur through various mechanisms [91–95]. The predominant mechanism recognized for its heating effect involves the generation of heat loss through induced eddy currents, particularly prevalent in bulk materials. Eddy current heating is attributed to induced currents, where electrons traverse the electrical resistance of the material [94,96]. Due to the low electrical conductivity of MNPs, the eddy current heat is very weak; hence, they are unable to generate a considerable electrical voltage [96].

Other mechanisms to take into account for heat transfer are (I) hysteresis losses and (II) relaxation losses [97]. Ferromagnetic materials naturally form magnetic domains where magnetic moments align uniformly. When exposed to alternating positive and negative magnetic fields, the material exhibits nonlinear magnetization behavior, depicted by a distinctive pattern known as the hysteresis loop. In a hysteresis loop, two mechanisms contribute to hysteresis losses. The first involves the pinning and release of domain walls at irregularities within the material, while the second is the rotation of magnetic moments within domains [98].

In medical application, relaxation losses are more prominent, which are divided into two kinds: "Neel Losses" and "Brownian Losses" [99]. The prevailing relaxation mechanism in the magnetic behavior of colloidal suspensions is dictated by the properties of the NPs [100].

The dissipation of heat from magnetic particles occurs due to the delayed relaxation of their magnetic moments, either through internal rotation within the particle (Néel mechanism) or the rotation of the particle itself (Brownian mechanism). This phenomenon occurs when the particles are subjected to an AC magnetic field with reversal times that are briefer than the magnetic relaxation times of the particles [101], as shown in Fig. 7 [102]. For small MNPs, the primary factor causing heat dissipation is the relaxation losses due to a delay in magnetization relaxation [103].

The mechanical or Brownian mechanism refers to the actual rotation of MNPs, which is defined by the time relaxation parameter τ_B [104] which is

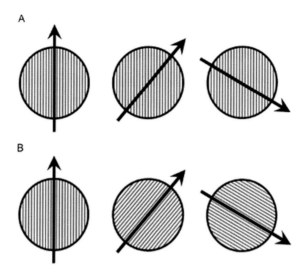

Fig. 7 Néel rotation vs. Brownian rotation. (A) Néel rotation: the magnetic moment rotates while the particle remains fixed. (B) Brownian rotation: the magnetic moment remains fixed with respect to the crystalline axes while the particle rotates [102].

$$\tau_B = \frac{3\eta V_{\text{hyd}}}{k_B T} \qquad (6)$$

Here, η represents the viscosity of the medium, V_{hyd} denotes the hydrodynamic volume of MNPs, and $k_B T$ corresponds to the thermal energy [104,105].

The Néel mechanism is effectively explained for non-interacting and monodomain MNPs through the Néel relaxation time, denoted as τ_N, as expressed by [104].

$$\tau_N = \tau_0 \exp\left(K_{\text{eff}} V_{\text{mag}} / k_B T\right) \qquad (7)$$

In this context K_{eff} and V_{mag} represent the effective anisotropy and magnetic volume of individual single-domain NPs, respectively [104,105]. Additionally, τ_0 represents the system's relaxation time, usually ranging between 10^{-9} to 10^{-11} seconds [104,105] However, in most research works, it is commonly considered as 10^{-9} seconds [105].

The above equations demonstrate that the Néel relaxation time is significantly impacted by nanoparticle size, whereas the Brownian relaxation time is influenced by both the hydrodynamic size of the nanoparticle and the viscosity of the fluid. This leads to the conclusion that smaller

MNPs are more suitable for intracellular hyperthermia, as they require less energy for their magnetic moments to rotate and encounter fewer restrictions on rotation within highly viscous cellular environments [106].

Magnetization reversal occurs through the process with the quicker relaxation time, and the combined relaxation time is determined by [107].

$$\frac{1}{\tau} = \frac{1}{\tau_B} + \frac{1}{\tau_N} \tag{8}$$

Heat is generated from MNPs due to a delay in the relaxation of the magnetic moment when subjected to an external AMF, occurring within a timeframe shorter than the NPs' relaxation times for magnetic reversal. This heat dissipation is quantified by the following equation [106,108].

$$P = \mu_0 X'' fH^2$$

In this context, P represents heat dissipation, μ_0 stands for the permeability of free space (also known as the magnetic field constant), X denotes the magnetic susceptibility, f signifies the frequency of the applied magnetic field, and H indicates the strength of the applied magnetic field [106,108].

3.7 Specific absorption rate

The thermal efficacy of MNPs is measured by the rate at which heat is dispersed per unit mass of the NPs, known as the specific absorption rate (SAR) [109]. SAR is the amount of heating power P, which is measured in W, produced for each gram of the nanoparticle's mass [110].

$$SAR = P/m_{MNP} \tag{9}$$

The size of particles and the specific surfactant employed in ferrite materials significantly impact the SAR value [111]. Higher power dissipation results in a reduction of the required concentrations of injected MNPs [112]. Table 2 [111,113] displays the correlation between SAR, the size of the NPs, and their corona conducted in studies.

3.8 Governing equations

In recent years, from initial investigations into whole-body hyperthermia to more recent applications aimed at targeting specific tumor masses, mathematical and computational modeling have played a crucial role in comprehending the underlying physics of hyperthermia [61]. Jiang et al. [46] have created a specialized computational framework that merges the lattice-Boltzmann method (LBM) modeling with the particle-swarm optimization (PSO) algorithm to

Table 2 Comparative SAR analysis across commonly used materials [111,113].

Magnetic component References	Core diameter (nm)	Size (nm)	Corona	H (kA/m)		ν(kHz)	SAR (W/gFe)
Magnetite	13	–	Aminosilan	13	520	146	[138]
Magnetite	18	–	Dextran	16	55	57	[139]
Magnetite	10	–	–	6.5	400	211[a]	[140]
Manganese ferrite	10	–	Liposomes	15	300	135	[141]
Cobalt ferrite	6	–	Mercaptoundecanoic acid	50	266	6[a]	[142]
Cobalt ferrite	18	–	Suspended in gel of agarose	30	108	3.5[a]	[143]
IONs	–	52 ± 16	Dextran-coated	58	292	81 ± 19	[113]
IONs	–	56 ± 11	Epichlorohydrin crosslinked dextran-coated	58	292	98 ± 16	[113]
IONs	–	52 ± 8	Amine-conjugated crosslinked dextran-coated	58	292	89 ± 9	[113]
IONs	–	–	CREKA-conjugated dextran-coated	58	292	66 ± 18	[113]

[a]$(W/g_{ferrite})$.

optimize the injection strategies of MNPs for hyperthermia-based cancer treatment. They applied this framework to two basic tumor models, one circular and the other elliptical as shown in Fig. 8, to gain insights into how the heat is distributed within the tumor and the surrounding healthy tissue and the subsequent temperature distribution. The findings have revealed that multi-site injection strategies are generally effective, while the single-site injection approach is ineffective, even for the simplest circular tumor model [46].

The mathematical equations that describe this multi-physics problem, where a simplified circular tumor is positioned at the center of a square block, are as follows:

$$\nabla \cdot \mathbf{u} = 0 \tag{10}$$

$$\frac{\partial \mathbf{u}}{\partial t} + (\mathbf{u} \cdot \nabla)\left(\frac{\mathbf{u}}{\varnothing}\right) = -\frac{1}{\rho_{nf}}\nabla(\varnothing p) + v_{nf}\nabla^2 \mathbf{u} + \frac{\mathbf{F}}{\rho_{nf}} \tag{11}$$

$$\sigma\frac{\partial T}{\partial t} + \mathbf{u} \cdot \nabla T = \alpha_e \nabla^2 T + \frac{1}{(pc_p)}\dot{m}_b c_{pb}(T_b - T) + \frac{Q}{(pc_p)_{nf}} \tag{12}$$

$$\varnothing\frac{\partial C}{\partial t} + \mathbf{u} \cdot \nabla C = D_e \nabla^2 C \tag{13}$$

Where

$$\nabla \equiv \frac{\partial}{\partial x}i + \frac{\partial}{\partial y}j \tag{14}$$

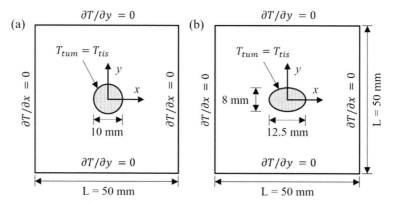

Fig. 8 Schematics of two tumor models surrounded by healthy tissue: (A) a circular tumor model, and (B) an elliptical tumor model. The gray regions represent the tumor, and the outer white regions represent the healthy tissue [46].

Specifically, these equations encompass the continuity equation, the momentum equation, the energy equation, and the concentration equation [52].

3.9 The impact of blood vessels on temperature patterns

Scientists, doctors, mathematicians, and engineers are working on developing relevant models to understand how heat moves in our bodies. This is important for developing new medical technology to treat serious illnesses like cancer, because we need to know how blood vessels and tissues interact with heat [114].

Ensuring consistent spreading of the imposed heat and the subsequent temperature distribution is vital in hyperthermia treatment, but achieving this, apart from whole-body therapy, proves challenging [115]. One significant cause of uneven temperature distribution could be the nearby blood vessel, which has a notable cooling effect on the temperature distribution in its presence [52], the other reason would be insufficient concentration of MNPs in the area [111].

According to Huang et al. [50] works, the dimensions of blood vessels within the vascular system don't notably impact the overall energy usage in hyperthermia therapy as long as the total rate of blood flow remains constant.

To reduce the impact of blood vessel on temperature variations, it is crucial to understand and manage the thermal effects of blood vessels, especially within the specific targeted [116]. The method employed by Huang et al. [117] involved a preheating area and adaptable optimization, which can successfully minimize the cooling impact caused by thermally significant blood vessels during hyperthermia therapy, resulting in a more consistent and therapeutic temperature within the tumor area.

4. Conclusions

Overall, hyperthermia is a promising approach in the treatment of cardiovascular disease and cancer. Ongoing research continues to explore its potential benefits and refine its applications. By selectively applying heat to specific areas of the body or the body as a whole, hyperthermia offers a multifaceted approach to addressing these serious medical conditions, potentially improving patient outcomes and quality of life.

Understanding how temperature influences LDL accumulation at the boundary between the endothelium and intima facilitates enhanced understanding of the potential mechanisms for atherosclerotic plaque development. External heating, instigating a higher intima temperature, appears to encourage LDL accumulation due to the Ludwig-Soret effect, while internal heating leads to a comparatively cooler intima temperature and reduced LDL concentration. Understanding the intricate temperature-driven mechanisms affecting LDL accumulation holds the potential for therapeutic interventions, though translating these discoveries into clinical applications requires further research.

In cancer treatment, nanoparticle-based hyperthermia offers targeted therapy prospects. Utilizing NPs like IONPs, AuNPs, and CNPs, represents a promising development in cancer treatment. Furthermore, advancements in managing thermal effects, such as minimizing the impact of blood vessels through optimization methods, offers a more consistent and therapeutic approach in hyperthermia treatment.

In essence, hyperthermia emerges as a compelling advancement in both CVD and cancer treatment. Its precision, guided by a deeper comprehension of temperature-based processes, opens entries for innovative therapies that could significantly enhance patient outcomes. However, effectively translating these insights into clinical practice demands continuous interdisciplinary research and technological advancements. The collaborative efforts among scientists, clinicians, and technology developers drive the exploration of novel areas in this domain, offering potential for enhanced patient outcomes and improved quality of life across diverse medical fields.

References

[1] G.M. Hahn, G.M. Hahn, Introduction historical and general comments, Hyperth. Cancer (1982) 1–5.

[2] H.R. Moyer, K.A. Delman, The role of hyperthermia in optimizing tumor response to regional therapy, Int. J. Hyperth. 24 (2008) 251–261.

[3] D. Sardari, N. Verga, Cancer treatment with hyperthermia, Citeseer (2011).

[4] P. Gas, Essential facts on the history of hyperthermia and their connections with electromedicine, ArXiv Prepr. ArXiv (2017) 171000652.

[5] G.C.R. Consortium, Global effect of modifiable risk factors on cardiovascular disease and mortality, N. Engl. J. Med. 389 (2023) 1273–1285.

[6] J.-E. Tarride, M. Lim, M. DesMeules, W. Luo, N. Burke, D. O'Reilly, et al., A review of the cost of cardiovascular disease, Can. J. Cardiology 25 (2009) e195–e202.

[7] M. Nichols, N. Townsend, P. Scarborough, M. Rayner, Cardiovascular disease in Europe 2014: epidemiological update, Eur. Heart J. 35 (2014) 2950–2959.

[8] W. Hao, A. Friedman, The LDL-HDL profile determines the risk of atherosclerosis: a mathematical model, PLoS One 9 (2014) e90497.

[9] X. Liu, Y. Fan, X. Deng, Effect of the endothelial glycocalyx layer on arterial LDL transport under normal and high pressure, J. Theor. Biol. 283 (2011) 71–81, https://doi.org/10.1016/j.jtbi.2011.05.030

[10] S.A. Everson-Rose, T.T. Lewis, Psychosocial factors and cardiovascular diseases, Annu. Rev. Public. Health 26 (2005) 469–500.

[11] J.F. Keaney Jr., Atherosclerosis: from lesion formation to plaque activation and endothelial dysfunction, Mol. Asp. Med. 21 (2000) 99–166.

[12] J. Abraham, J. Stark, J. Gorman, E. Sparrow, R. Kohler, A model of drug deposition within artery walls, J. Med. Device 7 (2013) 020902.

[13] S. Chung, K. Vafai, Mechanobiology of low-density lipoprotein transport within an arterial wall-Impact of hyperthermia and coupling effects, J. Biomech. 47 (2014) 137–147, https://doi.org/10.1016/j.jbiomech.2013.09.030

[14] H. Zhang, N. Ji, X. Gong, S. Ni, Y. Wang, NEAT1/miR-140-3p/MAPK1 mediates the viability and survival of coronary endothelial cells and affects coronary atherosclerotic heart disease, Acta Biochim. Biophys. Sin. (Shanghai) 52 (2020) 967–974.

[15] L. Badimon, G. Vilahur, Thrombosis formation on atherosclerotic lesions and plaque rupture, J. Intern. Med. 276 (2014) 618–632.

[16] M. Iasiello, K. Vafai, A. Andreozzi, N. Bianco, F. Tavakkoli, Effects of external and internal hyperthermia on LDL transport and accumulation within an arterial wall in the presence of a stenosis, Ann. Biomed. Eng. 43 (2015) 1585–1599, https://doi.org/10.1007/s10439-014-1196-0

[17] M. Iasiello, K. Vafai, A. Andreozzi, N. Bianco, Boundary layer considerations in a multi-layer model for LDL accumulation, Comput. Methods Biomech. Biomed. Engin 21 (2018) 803–811.

[18] S. Wang, K. Vafai, Analysis of low density lipoprotein (LDL) transport within a curved artery, Ann. Biomed. Eng. 43 (2015) 1571–1584, https://doi.org/10.1007/s10439-014-1219-x

[19] D.K. Stangeby, C.R. Ethier, Computational analysis of coupled blood-wall arterial LDL transport, J. Biomech. Eng. 124 (2002) 1–8.

[20] C.J. Schwartz, A.J. Valente, E.A. Sprague, J.L. Kelley, R.M. Nerem, The pathogenesis of atherosclerosis: an overview, Clin. Cardiol. 14 (1991) 1–16.

[21] J.R. Stark, J.M. Gorman, E.M. Sparrow, J.P. Abraham, R.E. Kohler, Controlling the rate of penetration of a therapeutic drug into the wall of an artery by means of a pressurized balloon, J. Biomed. Sci. Eng. (2013).

[22] L.B. Nielsen, Transfer of low density lipoprotein into the arterial wall and risk of atherosclerosis, Atherosclerosis 123 (1996) 1–15.

[23] Vekic, Jelena, A. Zeljkovic, A.F.G. Cicero, A. Janez, A.P. Stoian, A. Sonmez, M. Rizzo, Atherosclerosis development and progression: the role of atherogenic small, dense LDL, Medicina 58 (2022) 299.

[24] F. Calara, P. Dimayuga, A. Niemann, J. Thyberg, U. Diczfalusy, J.L. Witztum, et al., An animal model to study local oxidation of LDL and its biological effects in the arterial wall, Arterioscler. Thromb. Vasc. Biol. 18 (1998) 884–893.

[25] N. Yang, K. Vafai, Modeling of low-density lipoprotein (LDL) transport in the artery-effects of hypertension, Int. J. Heat. Mass. Transf. 49 (2006) 850–867, https://doi.org/10.1016/j.ijheatmasstransfer.2005.09.019

[26] L. Ai, K. Vafai, A coupling model for macromolecule transport in a stenosed arterial wall, Int. J. Heat. Mass. Transf. 49 (2006) 1568–1591.

[27] M. Prosi, P. Zunino, K. Perktold, A. Quarteroni, Mathematical and numerical models for transfer of low-density lipoproteins through the arterial walls: a new methodology for the model set up with applications to the study of disturbed lumenal flow, J. Biomech. 38 (2005) 903–917.

[28] S. Chung, K. Vafai, Effect of the fluid–structure interactions on low-density lipoprotein transport within a multi-layered arterial wall, J. Biomech. 45 (2012) 371–381.

[29] A. Nematollahi, E. Shirani, I. Mirzaee, M.R. Sadeghi, Numerical simulation of LDL particles mass transport in human carotid artery under steady state conditions, Sci. Iran. 19 (2012) 519–524.

[30] K. Jesionek, A. Slapik, M. Kostur, Effect of hypertension on low-density lipoprotein transport within a multi-layered arterial wall: modelling consistent with experiments 2016.

[31] N. Sun, N.B. Wood, A.D. Hughes, S.A.M. Thom, X.Y. Xu, Fluid-wall modelling of mass transfer in an axisymmetric stenosis: effects of shear-dependent transport properties, Ann. Biomed. Eng. 34 (2006) 1119–1128.

[32] M. Roustaei, M.R. Nikmaneshi, B. Firoozabadi, Simulation of low density lipoprotein (LDL) permeation into multilayer coronary arterial wall: Interactive effects of wall shear stress and fluid-structure interaction in hypertension, J. Biomech. 67 (2018) 114–122.

[33] N. Yang, K. Vafai, Low-density lipoprotein (LDL) transport in an artery—a simplified analytical solution, Int. J. Heat. Mass. Transf. 51 (2008) 497–505, https://doi.org/10.1016/j.ijheatmasstransfer.2007.05.023

[34] M. Iasiello, K. Vafai, A. Andreozzi, N. Bianco, Boundary layer considerations in a multi-layer model for LDL accumulation, Comput. Methods Biomech. Biomed. Engin 21 (2018) 803–811, https://doi.org/10.1080/10255842.2018.1521963

[35] S. Chung, K. Vafai, Low-density lipoprotein transport within a multi-layered arterial wall—effect of the atherosclerotic plaque/stenosis, J. Biomech. 46 (2013) 574–585.

[36] M. Iasiello, K. Vafai, A. Andreozzi, N. Bianco, Analysis of non-Newtonian effects on low-density lipoprotein accumulation in an artery, J. Biomech. 49 (2016) 1437–1446.

[37] J.P. Abraham, J.M. Gorman, E.M. Sparrow, J.R. Stark, R.E. Kohler, A mass transfer model of temporal drug deposition in artery walls, Int. J. Heat. Mass. Transf. 58 (2013) 632–638.

[38] S. Wang, K. Vafai, Analysis of the effect of stent emplacement on LDL transport within an artery, Int. J. Heat. Mass. Transf. 64 (2013) 1031–1040, https://doi.org/10.1016/j.ijheatmasstransfer.2013.05.041

[39] M. Iasiello, C. Tucci, A. Andreozzi, N. Bianco, K. Vafai, Modeling LDL Accumulation within an Arterial Wall, Elsevier Inc,, 2022 https://doi.org/10.1016/b978-0-323-85740-6.00015-7.

[40] M. Iasiello, K. Vafai, A. Andreozzi, N. Bianco, Low-density lipoprotein transport through an arterial wall under hyperthermia and hypertension conditions—an analytical solution, J. Biomech. 49 (2016) 193–204, https://doi.org/10.1016/j.jbiomech.2015.12.015

[41] N. Yang, K. Vafai, Low-density lipoprotein (LDL) transport in an artery—a simplified analytical solution, Int. J. Heat. Mass. Transf. 51 (2008) 497–505, https://doi.org/10.1016/j.ijheatmasstransfer.2007.05.023

[42] H. Sung, J. Ferlay, R.L. Siegel, M. Laversanne, I. Soerjomataram, A. Jemal, et al., Global cancer statistics 2020: GLOBOCAN estimates of incidence and mortality worldwide for 36 cancers in 185 countries, CA Cancer J. Clin. 71 (2021) 209–249.

[43] R. Siegel, C. DeSantis, K. Virgo, K. Stein, A. Mariotto, T. Smith, et al., Cancer treatment and survivorship statistics, 2012, CA Cancer J. Clin. 62 (2012) 220–241.

[44] J. Zugazagoitia, C. Guedes, S. Ponce, I. Ferrer, S. Molina-Pinelo, L. Paz-Ares, Current challenges in cancer treatment, Clin. Ther. 38 (2016) 1551–1566.

[45] M. Chang, Z. Hou, M. Wang, C. Li, J. Lin, Recent advances in hyperthermia therapy-based synergistic immunotherapy, Adv. Mater. 33 (2021) 2004788.

[46] Q. Jiang, F. Ren, C. Wang, Z. Wang, G. Kefayati, S. Kenjeres, et al., On the magnetic nanoparticle injection strategy for hyperthermia treatment, Int. J. Mech. Sci. 235 (2022) 107707, , https://doi.org/10.1016/j.ijmecsci.2022.107707

[47] A. Etminan, A. Dahaghin, S. Emadiyanrazavi, M. Salimibani, R. Eivazzadeh-Keihan, M. Haghpanahi, et al., Simulation of heat transfer, mass transfer and tissue damage in magnetic nanoparticle hyperthermia with blood vessels, J. Therm. Biol. 110 (2022) 103371, https://doi.org/10.1016/j.jtherbio.2022.103371

[48] Z. Kayani, N. Islami, N. Behzadpour, N. Zahraie, S. Imanlou, P. Tamaddon, et al., Combating cancer by utilizing noble metallic nanostructures in combination with laser photothermal and x-ray radiotherapy, J. Drug. Deliv. Sci. Technol. 65 (2021) 102689.

[49] M. Arruebo, N. Vilaboa, B. Sáez-Gutierrez, J. Lambea, A. Tres, M. Valladares, et al., Assessment of the evolution of cancer treatment therapies, Cancers (Basel) 3 (2011) 3279–3330.

[50] H.-W. Huang, T.-C. Shih, C.-T. Liauh, Predicting effects of blood flow rate and size of vessels in a vasculature on hyperthermia treatments using computer simulation, Biomed. Eng. Online 9 (2010) 1–19.

[51] S. Mahjoob, K. Vafai, Analytical characterization of heat transport through biological media incorporating hyperthermia treatment, Int. J. Heat. Mass. Transf. 52 (2009) 1608–1618, https://doi.org/10.1016/j.ijheatmasstransfer.2008.07.038

[52] Q. Jiang, F. Ren, C. Wang, Z. Wang, G. Kefayati, S. Kenjeres, et al., Simulation of magnetic hyperthermia cancer treatment near a blood vessel (2023) 1–36.

[53] M. Harabech, J. Leliaert, A. Coene, G. Crevecoeur, D. Van Roost, L. Dupré, The effect of the magnetic nanoparticle's size dependence of the relaxation time constant on the specific loss power of magnetic nanoparticle hyperthermia, J. Magn. Magn Mater. 426 (2017) 206–210.

[54] M.W. Dewhirst, J. Abraham, B. Viglianti, Evolution of thermal dosimetry for application of hyperthermia to treat cancer, Advanced Heat Transfer vol. 47, Elsevier,, 2015, pp. 397–421.

[55] A. Makridis, S. Curto, G.C. Van Rhoon, T. Samaras, M. Angelakeris, A standardisation protocol for accurate evaluation of specific loss power in magnetic hyperthermia, J. Phys. D. Appl. Phys. 52 (2019) 255001.

[56] A. Curcio, A.K.A. Silva, S. Cabana, A. Espinosa, B. Baptiste, N. Menguy, et al., Iron oxide nanoflowers@ CuS hybrids for cancer tri-therapy: interplay of photothermal therapy, magnetic hyperthermia and photodynamic therapy, Theranostics 9 (2019) 1288.

[57] P. Cherukuri, E.S. Glazer, S.A. Curley, Targeted hyperthermia using metal nanoparticles, Adv. Drug. Deliv. Rev. 62 (2010) 339–345, https://doi.org/10.1016/j.addr.2009.11.006

[58] R.M. Patil, N.D. Thorat, P.B. Shete, S.V. Otari, B.M. Tiwale, S.H. Pawar, In vitro hyperthermia with improved colloidal stability and enhanced SAR of magnetic core/ shell nanostructures, Mater. Sci. Eng.: C 59 (2016) 702–709.

[59] A. Włodarczyk, S. Gorgoń, A. Radoń, K. Bajdak-Rusinek, Magnetite nanoparticles in magnetic hyperthermia and cancer therapies: challenges and perspectives, Nanomaterials 12 (2022) 1–23, https://doi.org/10.3390/nano12111807

[60] J. Qu, G. Liu, Y. Wang, R. Hong, Preparation of Fe3O4–chitosan nanoparticles used for hyperthermia, Adv. Powder Technol. 21 (2010) 461–467.

[61] M. Nabil, P. Decuzzi, P. Zunino, Modelling mass and heat transfer in nano-based cancer hyperthermia, R. Soc. Open. Sci. 2 (2015), https://doi.org/10.1098/rsos.150447

[62] S. Xiao, Y. Tang, Z. Lv, Y. Lin, L. Chen, Nanomedicine–advantages for their use in rheumatoid arthritis theranostics, J. Control. Release 316 (2019) 302–316.

[63] S. Majee, G.C. Shit, Modeling and simulation of blood flow with magnetic nanoparticles as carrier for targeted drug delivery in the stenosed artery, Eur. J. Mech.-B/ Fluids 83 (2020) 42–57.

[64] J. Chomoucka, J. Drbohlavova, D. Huska, V. Adam, R. Kizek, J. Hubalek, Magnetic nanoparticles and targeted drug delivering, Pharmacol. Res. 62 (2010) 144–149.

[65] H.-W. Yang, M.-Y. Hua, H.-L. Liu, C.-Y. Huang, K.-C. Wei, Potential of magnetic nanoparticles for targeted drug delivery, Nanotechnol. Sci. Appl. (2012) 73–86.

[66] A. Avasthi, C. Caro, E. Pozo-Torres, M.P. Leal, M.L. García-Martín, Magnetic nanoparticles as MRI contrast agents, Surface-modified Nanobiomaterials for Electrochemical and Biomedicine Applications (2020) 49–91.

[67] C. Rümenapp, B. Gleich, A. Haase, Magnetic nanoparticles in magnetic resonance imaging and diagnostics, Pharm. Res. 29 (2012) 1165–1179.

[68] K. Khanafer, K. Vafai, A. Kangarlu, Water diffusion in biomedical systems as related to magnetic resonance imaging, Magn. Reson. Imaging 21 (2003) 17–31.

[69] K. Khanafer, K. Vafai, A. Kangarlu, Computational modeling of cerebral diffusion-application to stroke imaging, Magn. Reson. Imaging 21 (2003) 651–661.

[70] M.V. Yigit, A. Moore, Z. Medarova, Magnetic nanoparticles for cancer diagnosis and therapy, Pharm. Res. 29 (2012) 1180–1188.

[71] H. Fatima, K.-S. Kim, Magnetic nanoparticles for bioseparation, Korean J. Chem. Eng. 34 (2017) 589–599.

[72] I.J. Bruce, T. Sen, Surface modification of magnetic nanoparticles with alkoxysilanes and their application in magnetic bioseparations, Langmuir 21 (2005) 7029–7035.

[73] E. Kosari, K. Vafai, Transport and dynamic analysis of magnetic nanoparticles in brain microvascular vessels, Phys. Fluids 33 (2021).

[74] L. Kafrouni, O. Savadogo, Recent progress on magnetic nanoparticles for magnetic hyperthermia, Prog. Biomater. 5 (2016) 147–160.

[75] J. Jose, R. Kumar, S. Harilal, G.E. Mathew, D.G.T. Parambi, A. Prabhu, et al., Magnetic nanoparticles for hyperthermia in cancer treatment: an emerging tool, Environ. Sci. Pollut. Res. 27 (2020) 19214–19225.

[76] M. Bañobre-López, A. Teijeiro, J. Rivas, Magnetic nanoparticle-based hyperthermia for cancer treatment, Rep. Practical Oncol. Radiotherapy 18 (2013) 397–400, https://doi.org/10.1016/j.rpor.2013.09.011

[77] X. Sun, Y. Huang, D.E. Nikles, FePt and CoPt magnetic nanoparticles film for future high density data storage media, Int. J. Nanotechnol. 1 (2004) 328–346.

[78] M.E. de Sousa, M.B. Fernandez van Raap, P.C. Rivas, P. Mendoza Zélis, P. Girardin, G.A. Pasquevich, et al., Stability and relaxation mechanisms of citric acid coated magnetite nanoparticles for magnetic hyperthermia, J. Phys. Chem. C. 117 (2013) 5436–5445.

[79] C. Pucci, A. Degl'Innocenti, M.B. Gümüş, G. Ciofani, Superparamagnetic iron oxide nanoparticles for magnetic hyperthermia: Recent advancements, molecular effects, and future directions in the omics era, Biomater. Sci. 10 (2022) 2103–2121.

[80] Q. Jiang, F. Ren, C. Wang, Z. Wang, G. Kefayati, S. Kenjeres et al. Simulation of tumor ablation in hyperthermia cancer treatment: a parametric study 2023.

[81] R. Eivazzadeh-Keihan, F. Radinekiyan, A. Maleki, M.S. Bani, Z. Hajizadeh, S. Asgharnasl, A novel biocompatible core-shell magnetic nanocomposite based on cross-linked chitosan hydrogels for in vitro hyperthermia of cancer therapy, Int. J. Biol. Macromol. 140 (2019) 407–414.

[82] D.K. Mondal, G. Phukan, N. Paul, J.P. Borah, Improved self heating and optical properties of bifunctional Fe_3O_4/ZnS nanocomposites for magnetic hyperthermia application, J. Magn. Magn Mater. 528 (2021) 167809.

[83] K.C. Barick, S. Singh, D. Bahadur, M.A. Lawande, D.P. Patkar, P.A. Hassan, Carboxyl decorated Fe_3O_4 nanoparticles for MRI diagnosis and localized hyperthermia, J. Colloid Interface Sci. 418 (2014) 120–125.

[84] S. He, H. Zhang, Y. Liu, F. Sun, X. Yu, X. Li, et al., Maximizing specific loss power for magnetic hyperthermia by hard–soft mixed ferrites, Small 14 (2018) 1800135.

[85] R. Hergt, S. Dutz, M. Röder, Effects of size distribution on hysteresis losses of magnetic nanoparticles for hyperthermia, J. Phys.: Condens. Matter 20 (2008) 385214.

[86] Z. Li, M. Kawashita, N. Araki, M. Mitsumori, M. Hiraoka, M. Doi, Magnetite nanoparticles with high heating efficiencies for application in the hyperthermia of cancer, Mater. Sci. Eng.: C. 30 (2010) 990–996.

[87] T. Atsumi, Fundamental studies of hyperthermia using magnetic particles as thermoseeds 1: Development of magnetic particles suitable for hyperthermia, J. Magn. Soc. Jpn. 30 (2006) 555–560.

[88] M.R.M. Radzi, N.A. Johari, W.F.A.W.M. Zawawi, N.A. Zawawi, N.A. Latiff, N.A.N.N. Malek, et al., In vivo evaluation of oxidized multiwalled-carbon nanotubes-mediated hyperthermia treatment for breast cancer, Biomater. Adv. 134 (2022) 112586.

[89] Y. Shen, X. Zhang, L. Liang, J. Yue, D. Huang, W. Xu, et al., Mitochondria-targeting supra-carbon dots: Enhanced photothermal therapy selective to cancer cells and their hyperthermia molecular actions, Carbon N. Y. 156 (2020) 558–567.

[90] M.A. Behnam, F. Emami, Z. Sobhani, O. Koohi-Hosseinabadi, A.R. Dehghanian, S.M. Zebarjad, et al., Novel combination of silver nanoparticles and carbon nanotubes for plasmonic photo thermal therapy in melanoma cancer model, Adv. Pharm. Bull. 8 (2018) 49.

[91] J. Dobson, Magnetic nanoparticles for drug delivery, Drug. Dev. Res. 67 (2006) 55–60.

[92] Q.A. Pankhurst, J. Connolly, S.K. Jones, J.J.J. Dobson, Applications of magnetic nanoparticles in biomedicine, J. Phys. D. Appl. Phys. 36 (2003) R167.

[93] J.-H. Lee, J. Jang, J. Choi, S.H. Moon, S. Noh, J. Kim, et al., Exchange-coupled magnetic nanoparticles for efficient heat induction, Nat. Nanotechnol. 6 (2011) 418–422.

[94] R.V. Stigliano, F. Shubitidze, J.D. Petryk, L. Shoshiashvili, A.A. Petryk, P.J. Hoopes, Mitigation of eddy current heating during magnetic nanoparticle hyperthermia therapy, Int. J. Hyperth. 32 (2016) 735–748.

[95] R. Hergt, R. Hiergeist, M. Zeisberger, G. Glöckl, W. Weitschies, L.P. Ramirez, et al., Enhancement of AC-losses of magnetic nanoparticles for heating applications, J. Magn. Magn Mater. 280 (2004) 358–368.

[96] Suriyanto, E.Y.K. Ng, S.D. Kumar, Physical mechanism and modeling of heat generation and transfer in magnetic fluid hyperthermia through Néelian and Brownian relaxation: a review, Biomed. Eng. Online 16 (2017) 1–22.

[97] A. Kuncser, N. Iacob, V.E. Kuncser, On the relaxation time of interacting super-paramagnetic nanoparticles and implications for magnetic fluid hyperthermia, Beilstein J. Nanotechnol. 10 (2019) 1280–1289.

[98] P. O'Brien, Nanoscience: Volume 1: Nanostructures through Chemistry, Royal Society of Chemistry, 2012.

[99] R. Hergt, R. Hiergeist, I. Hilger, W.A. Kaiser, Y. Lapatnikov, S. Margel, et al., Maghemite nanoparticles with very high AC-losses for application in RF-magnetic hyperthermia, J. Magn. Magn Mater. 270 (2004) 345–357.

[100] M. Osaci, M. Cacciola, About the influence of the colloidal magnetic nanoparticles coating on the specific loss power in magnetic hyperthermia, J. Magn. Magn Mater. 519 (2021) 167451.

[101] M. Suto, Y. Hirota, H. Mamiya, A. Fujita, R. Kasuya, K. Tohji, et al., Heat dissipation mechanism of magnetite nanoparticles in magnetic fluid hyperthermia, J. Magn. Magn Mater. 321 (2009) 1493–1496.

[102] A.E. Deatsch, B.A. Evans, Heating efficiency in magnetic nanoparticle hyperthermia, J. Magn. Magn Mater. 354 (2014) 163–172.

[103] K. Murase, A simulation study on the specific loss power in magnetic hyperthermia in the presence of a static magnetic field, Open. J. Appl. Sci. 6 (2016) 839–851.

[104] F. Fabris, J. Lohr, E. Lima, A.A. De Almeida, H.E. Troiani, L.M. Rodríguez, et al., Adjusting the Néel relaxation time of Fe_3O_4/Zn x $Co1-$ x Fe_2O_4 core/shell nanoparticles for optimal heat generation in magnetic hyperthermia, Nanotechnology 32 (2020) 065703.

[105] M. Cobianchi, A. Guerrini, M. Avolio, C. Innocenti, M. Corti, P. Arosio, et al., Experimental determination of the frequency and field dependence of Specific Loss Power in Magnetic Fluid Hyperthermia, J. Magn. Magn Mater. 444 (2017) 154–160.

[106] P. Das, M. Colombo, D. Prosperi, Recent advances in magnetic fluid hyperthermia for cancer therapy, Colloids Surf. B Biointerfaces 174 (2019) 42–55.

[107] L.-Y. Zhang, H.-C. Gu, X.-M. Wang, Magnetite ferrofluid with high specific absorption rate for application in hyperthermia, J. Magn. Magn Mater. 311 (2007) 228–233.

[108] M. Suto, Y. Hirota, H. Mamiya, A. Fujita, R. Kasuya, K. Tohji, et al., Heat dissipation mechanism of magnetite nanoparticles in magnetic fluid hyperthermia, J. Magn. Magn Mater. 321 (2009) 1493–1496.

[109] D. Bonvin, A. Arakcheeva, A. Millán, R. Pinol, H. Hofmann, M.M. Ebersold, Controlling structural and magnetic properties of IONPs by aqueous synthesis for improved hyperthermia, RSC Adv. 7 (2017) 13159–13170.

[110] R.R. Wildeboer, P. Southern, Q.A. Pankhurst, On the reliable measurement of specific absorption rates and intrinsic loss parameters in magnetic hyperthermia materials, J. Magn. Magn Mater. 47 (2014) 495003.

[111] I. Sharifi, H. Shokrollahi, S. Amiri, Ferrite-based magnetic nanofluids used in hyperthermia applications, J. Magn. Magn Mater. 324 (2012) 903–915.

[112] Suriyanto, E.Y.K. Ng, S.D. Kumar, Physical mechanism and modeling of heat generation and transfer in magnetic fluid hyperthermia through Néelian and Brownian relaxation: a review, Biomed. Eng. Online 16 (2017) 1–22.

[113] A.M. Kruse, S.A. Meenach, K.W. Anderson, J.Z. Hilt, Synthesis and characterization of CREKA-conjugated iron oxide nanoparticles for hyperthermia applications, Acta Biomater. 10 (2014) 2622–2629.

[114] P.K. Gupta, J. Singh, K.N. Rai, A numerical study on heat transfer in tissues during hyperthermia, Math. Comput. Model. 57 (2013) 1018–1037, https://doi.org/10.1016/j.mcm.2011.12.050

[115] J. Crezee, J.J.W. Lagendijk, Temperature uniformity during hyperthermia: the impact of large vessels, Phys. Med. Biol. 37 (1992) 1321.

[116] H.-W. Huang, C.-T. Liauh, T.-C. Shih, T.-L. Horng, W.-L. Lin, Significance of blood vessels in optimization of absorbed power and temperature distributions during hyperthermia, Int. J. Heat. Mass. Transf. 53 (2010) 5651–5662.

[117] H.-W. Huang, C.-T. Liauh, T.-L. Horng, T.-C. Shih, C.-F. Chiang, W.-L. Lin, Effective heating for tumors with thermally significant blood vessels during hyperthermia treatment, Appl. Therm. Eng. 50 (2013) 837–847.

[118] G.M. Sulaiman, H.M. Waheeb, M.S. Jabir, S.H. Khazaal, Y.H. Dewir, Y. Naidoo, Hesperidin loaded on gold nanoparticles as a drug delivery system for a successful biocompatible, anti-cancer, anti-inflammatory and phagocytosis inducer model, Sci. Rep. 10 (2020) 9362.

[119] Y. Yang, M. Chen, Y. Wu, P. Wang, Y. Zhao, W. Zhu, et al., Ultrasound assisted one-step synthesis of Au@ Pt dendritic nanoparticles with enhanced NIR absorption for photothermal cancer therapy, RSC Adv. 9 (2019) 28541–28547.

[120] S.-M. Lee, H.J. Kim, Y.-J. Ha, Y.N. Park, S.-K. Lee, Y.-B. Park, et al., Targeted chemo-photothermal treatments of rheumatoid arthritis using gold half-shell multifunctional nanoparticles, ACS Nano 7 (2013) 50–57.

[121] P.K. Pandey, R. Maheshwari, N. Raval, P. Gondaliya, K. Kalia, R.K. Tekade, Nanogold-core multifunctional dendrimer for pulsatile chemo-, photothermal-and photodynamic-therapy of rheumatoid arthritis, J. Colloid Interface Sci. 544 (2019) 61–77.

[122] J. Vonnemann, N. Beziere, C. Böttcher, S.B. Riese, C. Kuehne, J. Dernedde, et al., Polyglycerolsulfate functionalized gold nanorods as optoacoustic signal nanoamplifiers for in vivo bioimaging of rheumatoid arthritis, Theranostics 4 (2014) 629.

[123] S. Al Tamimi, S. Ashraf, T. Abdulrehman, A. Parray, S.A. Mansour, Y. Haik, et al., Synthesis and analysis of silver–copper alloy nanoparticles of different ratios manifest anticancer activity in breast cancer cells, Cancer Nanotechnol. 11 (2020) 1–16.

[124] S. Kang, K. Kang, A. Chae, Y.-K. Kim, H. Jang, D.-H. Min, Fucoidan-coated coral-like Pt nanoparticles for computed tomography-guided highly enhanced synergistic anticancer effect against drug-resistant breast cancer cells, Nanoscale 11 (2019) 15173–15183.

[125] V. Ramalingam, S. Raja, M. Harshavardhan, In situ one-step synthesis of polymer-functionalized palladium nanoparticles: an efficient anticancer agent against breast cancer, Dalton Trans. 49 (2020) 3510–3518.

[126] S. Bagdonas, G. Kirdaite, G. Streckyte, V. Graziene, L. Leonaviciene, R. Bradunaite, et al., Spectroscopic study of ALA-induced endogenous porphyrins in arthritic knee tissues: targeting rheumatoid arthritis PDT, Photochem. Photobiol. Sci. 4 (2005) 497–502.

[127] A. Nakamura, T. Osonoi, Y. Terauchi, Relationship between urinary sodium excretion and pioglitazone-induced edema, J. Diabetes Investig. 1 (2010) 208–211.

[128] Y. Li, T. Lin, Y. Luo, Q. Liu, W. Xiao, W. Guo, et al., A smart and versatile theranostic nanomedicine platform based on nanoporphyrin, Nat. Commun. 5 (2014) 4712.

[129] P. Sathishkumar, Z. Li, R. Govindan, R. Jayakumar, C. Wang, F.L. Gu, Zinc oxide-quercetin nanocomposite as a smart nano-drug delivery system: Molecular-level interaction studies, Appl. Surf. Sci. 536 (2021) 147741.

[130] S.L. Gawali, S.B. Shelar, J. Gupta, K.C. Barick, P.A. Hassan, Immobilization of protein on Fe_3O_4 nanoparticles for magnetic hyperthermia application, Int. J. Biol. Macromol. 166 (2021) 851–860.

[131] M. Sharifi, A. Hasan, N.M.Q. Nanakali, A. Salihi, F.A. Qadir, H.A. Muhammad, et al., Combined chemo-magnetic field-photothermal breast cancer therapy based on porous magnetite nanospheres, Sci. Rep. 10 (2020) 5925.

[132] N. Zhang, Y. Wang, C. Zhang, Y. Fan, D. Li, X. Cao, et al., LDH-stabilized ultrasmall iron oxide nanoparticles as a platform for hyaluronidase-promoted MR imaging and chemotherapy of tumors, Theranostics 10 (2020) 2791.

[133] M. Fournelle, W. Bost, I.H. Tarner, T. Lehmberg, E. Weiß, R. Lemor, et al., Antitumor necrosis factor-α antibody-coupled gold nanorods as nanoprobes for molecular optoacoustic imaging in arthritis, Nanomedicine 8 (2012) 346–354.

[134] Q. Tang, J. Cui, Z. Tian, J. Sun, Z. Wang, S. Chang, et al., Oxygen and indocyanine green loaded phase-transition nanoparticle-mediated photo-sonodynamic cytotoxic effects on rheumatoid arthritis fibroblast-like synoviocytes, Int. J. Nanomed. (2017) 381–393.

[135] B. Gong, Y. Shen, H. Li, X. Li, X. Huan, J. Zhou, et al., Thermo-responsive polymer encapsulated gold nanorods for single continuous wave laser-induced photodynamic/photothermal tumour therapy, J. Nanobiotechnology 19 (2021) 1–14.

[136] H. Wang, Y. Liang, Y. Yin, J. Zhang, W. Su, A.M. White, et al., Carbon nano-onion-mediated dual targeting of P-selectin and P-glycoprotein to overcome cancer drug resistance, Nat. Commun. 12 (2021) 312.

[137] K. Manivannan, C.-C. Cheng, R. Anbazhagan, H.-C. Tsai, J.-K. Chen, Fabrication of silver seeds and nanoparticle on core-shell Ag@ SiO2 nanohybrids for combined photothermal therapy and bioimaging, J. Colloid Interface Sci. 537 (2019) 604–614.

[138] A. Jordan, R. Scholz, P. Wust, H. Schirra, T. Schiestel, H. Schmidt, et al., Endocytosis of dextran and silan-coated magnetite nanoparticles and the effect of intracellular hyperthermia on human mammary carcinoma cells in vitro, J. Magn. Magn Mater. 194 (1999) 185–196.

[139] L.-Y. Zhang, H.-C. Gu, X.-M. Wang, Magnetite ferrofluid with high specific absorption rate for application in hyperthermia, J. Magn. Magn Mater. 311 (2007) 228–233.

[140] I. Hilger, W. Andra, R. Hergt, R. Hiergeist, H. Schubert, W.A. Kaiser, Electromagnetic heating of breast tumors in interventional radiology: in vitro and in vivo studies in human cadavers and mice, Radiology 218 (2001) 570–575.

[141] P. Pradhan, J. Giri, R. Banerjee, J. Bellare, D. Bahadur, Preparation and characterization of manganese ferrite-based magnetic liposomes for hyperthermia treatment of cancer, J. Magn. Magn Mater. 311 (2007) 208–215.

[142] D.-H. Kim, D.E. Nikles, D.T. Johnson, C.S. Brazel, Heat generation of aqueously dispersed $CoFe_2O_4$ nanoparticles as heating agents for magnetically activated drug delivery and hyperthermia, J. Magn. Magn Mater. 320 (2008) 2390–2396.

[143] M. Veverka, P. Veverka, O. Kaman, A. Lančok, K. Závěta, E. Pollert, et al., Magnetic heating by cobalt ferrite nanoparticles, Nanotechnology 18 (2007) 345704.

CHAPTER THREE

Enhancement of heat transfer with nanofluids and its applications in heat exchangers

**Wajahat Ahmed Khan[a], Kaleemullah Shaikh[b], Rab Nawaz[b],
Salim Newaz Kazi[b,*], and Mohd Nashrul Mohd Zubir[b]**

[a]Nanotechnology & Catalysis Research Centre (NANOCAT), Institute for Advanced Studies (IAS), Universiti Malaya, Kuala Lumpur, Malaysia
[b]Department of Mechanical Engineering, Faculty of Engineering, Universiti Malaya, Kuala Lumpur, Malaysia
*Corresponding author. e-mail address: salimnewaz@um.edu.my

Contents

1. Introduction	102
2. Application of nanofluids in solar collectors	105
2.1 Direct absorption solar collectors (DASC)	106
2.2 Flat plate solar collectors (FPSC)	107
2.3 Evacuated tube solar collectors (ETSC)	109
3. Application of nanofluids for fouling retardation and heat transfer enhancement	111
3.1 Impact of multiwall carbon nanotubes (MWCNT) based nanofluid on retardation of fouling and enhancement of heat transfer	115
3.2 Impact of graphene nanoplatelet-based nanofluid (GNP) on fouling retardation and heat transfer enhancement	117
4. Applications of nanofluids in annular flow heat exchangers	118
4.1 Nanofluids behavior in an annular flow heat exchanger	120
5. Conclusion	123
Acknowledgements	124
References	124

Abstract

Intended or unintended generation of heat takes place throughout all the industrial activities, making heat exchangers a ubiquitous part of industries in the modern world. One of the chief influences of operation of heat exchangers is the selection of heat transfer fluid. Conventional heat transfer fluids (water, oil, air/steam, molten salts, etc.) have limited inherent abilities to meet advanced engineering demands. One way to enhance thermal properties is by dispersion of nanoparticles into base fluids, called nanofluids. Nanoparticles may include different materials such as metallic, metallic

---

Advances in Heat Transfer, Volume 57
ISSN 0065-2717, https://doi.org/10.1016/bs.aiht.2024.05.001
Copyright © 2024 Elsevier Inc. All rights are reserved, including those for text and data mining, AI training, and similar technologies.

oxide, ceramic, or carbonaceous substances. When properly suspended in a base liquid, they affect the thermophysical characteristics of the base fluid thereby providing an opportunity to engineer the fluid for enhanced performance. Nanofluids have been extensively investigated under certain applications to enhance the performance of devices. In this chapter, the effects of different properties and their relative importance as engineering nanofluids are discussed. Moreover, nanofluid applications in three areas (solar collectors, fouling mitigation, and annular flow) were discussed. Nanofluids were observed to enhance thermal performance of non-concentrating solar collectors, mitigate mineral fouling and enhance heat transfer in annular flow passages. In general, it is observed that carbon-based nanofluids perform better than other nanomaterial-based nanofluids.

1. Introduction

Heat transfer liquids play a crucial function in heat exchangers and their characteristics affect the performance and longevity of heat exchangers. However conventional heat transfer liquids such as engine coolants, lubricants and water have inherently low thermal conductivity [1–3]. Hence many investigations have been performed to improve heat exchanger performance through enhancing heat transfer fluids (HTFs). Many investigations have concentrated on enhancing thermal efficiency through suspending solid nanoparticles in base fluids. Initially, the size of suspended particles was in the millimeter or micrometer range [4–6]. Later, Choi suggested that thermal conductivity can be improved through the addition of nano-sized particles into conventional liquids and coined the term 'nanofluid' [7]. Such suspensions often exhibit higher heat transfer performance and better rheological properties [8–10]. Utilizing sustainable nanofluids has a positive impact on the environment through decreased energy usage and minimized environmental effects during heat transfer processes. For optimal cost–effectiveness and affordability, it is crucial to conduct a comprehensive analysis of the economic viability for utilizing specific nanoparticle-based nanofluids. Therefore, stable suspension of nanoparticles in the base liquid is one of the preferred approaches to increase the overall efficiency in the system.

Nanomaterials are known for their significant attributes, including heat capacity, large interfacial area, and elevated thermal conductivity. These factors greatly enhance the thermal efficiency of heat exchangers. A goal of nanofluids is to maximize thermal conductivities with a minimal amount of nanoparticles. A common nanofluid generally contains solid nanoparticles at very low concentrations (preferably $<1\%$ by mass) and without any surfactants or dispersion agents. Thermophoresis, diffusiophoresis, and Brownian motion of nanofluids contribute to enhanced thermal properties [11,12].

The choice of nanofluids is influenced by their ability to increase heat transfer their sustainability, and whether non-metallic particles can be used. Numerous investigations have been reported in solar collectors and heat exchangers using a variety of ceramics and metals [13,14]. Because of the complexity of nanofluid systems, altering system performance needs knowledge of the essential variables and nanofluid characteristics.

Timofeeva et al. [15,16] organized the trends of nanoparticle suspensions in the literature using a basic decision matrix. Each cell in the table provided the strength of impact of specific variables to the characteristics of nanofluid with "×", "▲", "○", and "■" symbols representing no, weak, medium, and strong dependencies, respectively, with scores of 0.0, 0.25, 0.5, and 1.0. Based on the table, engineering variables can be ordered on the basis of the increasing significance for their performance: The parameters effects were quantified and rank ordered as: Kapitza resistance (2.0) < additives (2.75) < temperature (3.75) ≈ particle shape (3.75) < nanoparticle material (4.0) ≈ surface charge (4.0) < nanoparticle size (5.0) < base fluid (5.25) < particle concentration (6.25) [15,16] where the numbers represent the relative importance. Such quantification of parameters through relative importance, reduces subjectiveness and promotes objectivity while engineering the nanofluid. The concentration of nano additive, base liquid, and size of the particle seem to be extremely important criteria for increasing the heat transfer effectiveness of a nanofluid (Table 1).

Numerous methods have been established to synthesize nanomaterials such as microemulsion production, reverse micelle methods or precipitation approaches, and mechanical attrition. However, most nanomaterials utilized in nanofluids have been produced through physical production, such as the inert-gas-condensation (IGC) procedure, or chemical production approaches, i.e., chemical vapor deposition [17,18]. To prepare a stable nanofluid, the particles must be appropriately small to be suspended through Brownian motion, or the particles should be protected against agglomeration with protective coatings, electric charge, or another approach [19–21]. Because agglomeration of nanoparticles takes place easily at high concentrations of nanoparticles (>20% by mass), low concentration recommended. In this chapter three major applications of heat exchangers with nanofluids will be focused on, namely, solar collectors, fouling mitigation, and annular-flow heat exchangers.

Table 1 Symbols: ■ – strongly dependent, ○ – medium dependent, ▲ – weakly dependent, × – no dependent, ? – unknown relation or alters from system to system, ⟰ – larger the better, ⟱ – smaller the better, ↑ – increases with the characteristic, ↓ – reduces with the characteristic.

	Nanofluid parameters	Nanoparticle material	Nanoparticle concentration	Nanoparticle shape	Nanoparticle size	Base fluid	Zeta potential /fluid pH	Kapitza resistance	Additives	Temperature
Nanofluid Properties										
Stability	⟰	▲	▲	▲	■↓	○	■	×	■	?
Density	⟰	■	■↓	×	×	■	×	×	×	×
Specific Heat	⟰	■	■↓	×	×	■	×	×	×	▲
Thermal Conductivity	⟰	○	■↓	○	■↓	▲	○	■↓	▲	○
Viscosity	⟰	▲	■↓	■	■↓	■↓	■	×	○	■
Heat Transfer Coefficient	⟰	■	■↓	■	■↓	■	■	■↓	○	■
Pumping Power Penalty	⟰	×	■	■	■↓	■	■	×	○	■
Relative Importance		4.0	6.25	3.75	5.0	5.25	4.0	2.0	2.75	3.75

2. Application of nanofluids in solar collectors

Solar collectors transfer incoming irradiance from the sun into useful thermal energy. The energy can either be stored or transferred to a working fluid for further utilization. Because of the increased demand of energy and shifts towards renewable energy, the use of solar collectors have been rising for the past two decades. There are a variety of collectors that have been designed for enhanced capture and conversion of solar energy. They are generally classified as concentrating and non-concentrating collectors or high-, medium- and low-temperature collectors. Flat plate solar collectors (FPSC) are by far the most generally utilized low temperature (<100°C) and inexpensive collector. With the availability of enhanced coatings, advanced FPSC's stagnation temperature can be raised to > 200°C. Other common types of non-concentrating collector is the evacuated tube solar collector (ETSC) and direct absorption solar collector (DASC) [22,23].

These solar collectors can also termed direct or indirect (open loop or close loop systems) — depending on the utilization of heated fluid, if heat is directly utilized (direct) or transferred or stored (indirect); and whether active or passive (natural or forced circulation) is used. Passive systems are typically less expensive than active systems, but they are also less effective. Passive systems rely on the buoyant flow of a heating medium. The density of water decreases with increasing temperature, causing the hot water to rise to the top of the collector, where it is stored in the tank. Fluid recirculates to the collector as it cools in the lower section of the tank.

The performance of FPSC, ETSC, and DASC systems is influenced by various parameters like material composition, absorber plate coating, shape, the number of tubes, the type of glaze, spacing among tubes, and insulation. The collector's efficiency can also be influenced through factors like atmospheric temperature, wind speed, solar irradiance, and flow rate. Lots of research focusing these parameter for improving solar collectors have been published [24]. One important component of solar collectors is the absorber surface. The surfaces used to absorb incoming solar energy should ideally have high absorptance (α) and low emittance (ε) values to increase solar radiation absorption and reduce radiation emission. Materials with high values of α_s/ε, like clean galvanized sheet metal with $\alpha_s/\varepsilon = 5.0$, are preferred for efficient heat collection. Materials with low values of α_s/ε, like anodized aluminum with $\alpha_s/\varepsilon = 0.17$, are preferred for heat rejection [25].

The most common fluids utilized in these collectors are water, glycols, and oils. Despite the fact that pure substances and regularly used fluids seldom fulfill all practical and performance requirements, the development of novel solar heat transfer fluids remains an expanding field. Many reviews have been published highlighting its progress [26–32], some of the latest investigations of nanofluids are discussed below.

2.1 Direct absorption solar collectors (DASC)

Recently, Wang et al. [33] tackled the issue of effectively harnessing solar energy through integrating nanofluids into solar collectors and capitalized on the distinct characteristics of 2D nanomaterials. Their study concentrated on MXene nanosheets produced via in-situ etching and then compared them with graphene nanofluid. Investigation revealed that MXene nanofluid has superior optical qualities due to specified surface plasmon quality, but the graphene based nanofluid showed greater thermal conductivity. The nanofluid of MXene achieved a 63.35% efficiency of heat-to-energy conversion at a 20 ppm concentration, which was 4.34% higher than graphene nanofluid. In 2022, Sadegh, Hosseini, and Dehaj [34] performed an investigation on the utilization of nanofluids comprising Fe_2O_3 and Fe_3O_4 nanoparticles at varying concentrations (0.02%, 0.01%, and 0.005%). The study found that despite having similar sizes and shapes, iron oxide nanoparticles exhibit different behaviors when dissolved in water. The Fe_2O_3 nanofluid exhibited significant absorption in the visible light spectrum whereas the Fe_3O_4 nanofluid showed lower absorption, primarily in the ultraviolet spectrum. It was discovered that the 0.02% Fe_2O_3 nanofluid absorbed all sunlight within 4 cm of the surface. However, nanofluid with 0.02% Fe_3O_4 only managed to absorb around 83% of the sunlight at an equivalent depth. The Fe_2O_3 nanofluid outperformed the Fe_3O_4 nanofluid at increasing solar-to-thermal energy conversion, achieving a maximum efficiency of 73% with 0.02% Fe_2O_3 nanofluid compared to 51% with Fe_3O_4. These results are displayed in Fig. 1, where Ti and Ta is inlet and

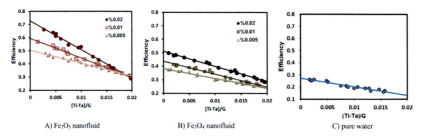

Fig. 1 The efficiency with normalized difference temperature; (A) Fe_2O_3 nanofluids, (B) Fe_3O_4 nanofluids, and (C) water.

ambient temperatures respectively and G is incident solar irradiance. Figure also highlights that the efficiency for water, Fe_3O_4 (0.005 wt%) and Fe_2O_3 (0.005 wt%) is around 30%, 40% and 50% respectively. Their research emphasizes that Fe_2O_3 nanofluids showed promise in enhancing the effectiveness of solar-energy conversion in DASC, yielding a viable choice for increasing use of solar energy.

2.2 Flat plate solar collectors (FPSC)

Hussein et al. [35] experimentally examined the efficiency of hybrid nanofluids in a FPSC. The hybrid nanofluid consisted of 20% covalent functionalized (CF)-MWCNTs + 20% CF-GNPs and 60% h-BN with water as a working fluid, both at the ratio of 40:60% in 0.1 wt% were made through a 2-step procedure. Their results at concentrations of 0.05, 0.08 and 0.1 (mass concentrations) indicated that the hybrid nanofluid enhanced thermal efficacy, reaching 85% (approximately 20% greater than the DW) at 0.1%. Also, at 20 °C and 60 °C, the thermal conductivity was raised by 12.0% and 64%, respectively. The increased mass concentration of nano additives and temperature in the hybrid nanofluid led to this outcome. The hybrid nanofluid exhibited its best performance at a concentration of 0.1%. The useful thermal power (Q_u) absorbed through the collector is demonstrated in Fig. 2A in comparison with the temperature difference between the ambient air and the collector inlet as a driving force. The Q exhibited a trend similar to solar radiation because it is mostly impacted through incoming irradiation.

Moradi et al. [36] experimentally explored the effect of nanoparticle temperature and concentration on the thermal conductivity of a hybrid

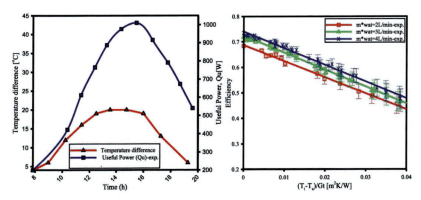

Fig. 2 (A) Useful power (W) and temperature difference (°C) versus time (h); (B) Collector efficiency with the different flow rate.

nanofluid (TiO$_2$/MWCNTs dispersed in EG/water). The investigation was performed at numerous temperatures varying from 20 to 60 °C and concentrations varying from 0.0625% to 1.0%, as shown in Fig. 3. The experimental outcomes revealed that thermal conductivity rose with an increase in temperature and nanoparticle concentration.

The Mouromtseff number (Mo) is utilized to compare the relative heat transfer capabilities of numerous liquids. For fully developed internal laminar flow, the Nusselt number remains constant under the conditions of constant wall temperature and constant heat flux. Hence, the Mo number ratio is equivalent to the ratio of the thermal conductivity of the nanofluid to the thermal conductivity of the base liquid.

Li et al. [37] assessed the optical absorption, stability characteristics, and thermal conductivity of MWCNT/SiC based hybrid nanofluid. The study demonstrated a substantial improvement in both stability and solar energy absorption efficiency, reaching 99.90% at a thickness of 10 mm with a mass concentration of 0.5% of MWCNTs/SiC. The hybrid MWCNTs/SiC, with a concentration of 0.5% yielded exceptional absorption performance, absorbing 99.90% of solar energy near-infrared wavelength bands. Additionally, the conversion of solar thermal efficiency increased with the rise in the weight percentage.

The performance of nanofluid based solar collector is highly affected by the temperature variation of nanofluids and the duration of solar radiation. The temperature of all liquids increased with increased time of radiation. The rise in temperature and the highest equilibrium temperature of the nanofluids were greater than with basic EG. As an example, hybrid nanofluids with a concentration of 0.1 wt% exhibited the greatest differential temperature that was

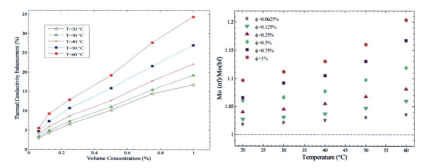

Fig. 3 (A) Hybrid nanofluid thermal conductivity enhancement against volume concentration at various temperatures and (B) efficiency of heat transfer for internal turbulent flow regime based on the Mouromtseff number.

107.1 °C after being exposed to radiation for 50 min. The rate of temperature rise was 2.14 °C/min. It was around 31.50% quicker than the base liquid. From the outcomes, the maximum values of η occurred after 10 mins of solar radiation for every concentration of nanofluids. This behavior was due to the rise in heat loss during the increase in temperature. It is important to understand that though similar intensities of radiation were utilized by different researchers, the efficiency of solar-thermal conversion was not similar due to the variation in receiver height. In the current investigation, the maximum efficiency was 97.3% at 1 wt% concentration. After 10 min, the efficiency declined until it feel to 72.60% after one hour, as illustrated in Fig. 4.

2.3 Evacuated tube solar collectors (ETSC)

Liang et al. [38] compared the thermal efficiency of three types of U-pipe ETSC devices both theoretically and experimentally. They found that the ETSC with three U-pipes reaches an efficiency limit of 0.82, significantly surpassing other types with equivalent heat exchange areas. Korres et al. [39] performed a CFD analysis and an experimental study on a U-type ETSC array. Predictions were made for thermal and exergy efficiencies. The deviation in thermal efficiency between the analytical solution and simulations results was found to be ~6%. Moreover, thermal efficiency was found to decline from the first module to the last because of higher inlet temperatures, implying higher thermal losses. Furthermore, they found that exergy increases as flow rate decreases because of the increase in fluid temperature.

Ersöz et al. [40] experimentally investigated the impacts of six working liquids (methanol, hexane, chloroform, petroleum ether, ethanol, and acetone) on a thermosyphon heat pipe evacuated-tube collector (THPETC)

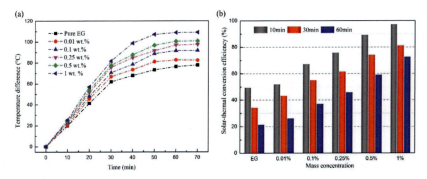

Fig. 4 (A) Temperature difference of SiC-MWCNTs nanofluids as a function of irradiation duration; (B) The solar-thermal conversion efficiency of SiC-MWCNTs nanofluids at variable irradiation times.

for air heating at velocities of 2, 3, and 4 m/s. The schematic of test rig is illustrated in Fig. 5A. Working liquids were charged in the thermosyphon heat pipe (THP) and then solar radiation, air velocity, inlet and outlet temperatures, and ambient temperature were measured. The findings indicated that acetone, at air velocities of 2.0 and 3.0 m/s, and chloroform, at an air velocity of 4 m/s yielded the greatest energy efficacy. The highest exergy efficiency was attained for acetone at an air velocity of 2 m/s, and for chloroform at air velocities of 3 and 4 m/s, as exhibited in Fig. 5.

Sadeghi et al. [41] built an experimental setup featuring an ETSC with a parabolic concentrator. Nanofluids comprised of Cu_2O/DW were employed as the heat transporting medium. The study found that higher volumetric flow rate and concentration of Cu_2O nanoparticles enhanced the thermal effectiveness of the ETSC, as shown in Fig. 6.

Mehmood et al. [42] modeled an ETSC water heating system with backup natural gas (composite system) using TRNSYS software. They reported that the composite system can annually save 23.0% to 56.0% of natural gas, that will decrease pollution emissions. Fig. 7 shows a system diagram of the TRNSYS simulation model.

In another study, Selvakumar et al. [43] investigated the thermal efficiency of an ETSC combined with a parabolic trough system employing therminol D-12 as the HTF. The system can attain temperatures as high as 68 °C with minimal solar radiation. Chopra et al. [44] investigated the performance of an ETSC scheme with phase change materials, the

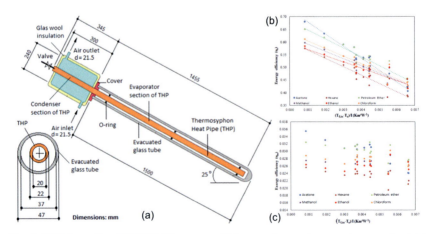

Fig. 5 (A) Schematic of THPETC (B) energy efficiency values vs temperature gradient (at air velocity 2 m/s), (C) exergy efficiencies vs temperature gradient (at air velocity 2 m/s).

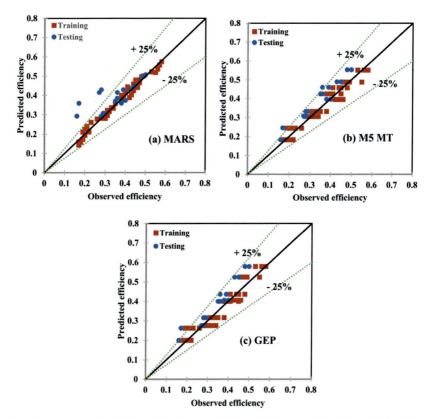

Fig. 6 Recommended AI models for predicting efficiency values. *From M.A. Ersöz, Effects of different working fluid use on the energy and exergy performance for evacuated tube solar collector with thermosyphon heat pipe, Renew. Energy 96 (2016) 244–256. https://doi.org/10.1016/j.renene.2016.04.058.*

schematic is shown in Fig. 8. The device's energy performance was found to be between 52.0–63.0% for a full day charge and 55.0–73.0% for midday charging. The peak thermal efficiency reached 72.52% during midday charging with a mass flow rate of 24.0 LPH. The PCM achieved an efficiency ranging from 61% to 64% for both charging modes.

3. Application of nanofluids for fouling retardation and heat transfer enhancement

Accumulation and removal of deposits on the surface of heat exchanger is called fouling. Fig. 9 exhibits fouling on the surface of a

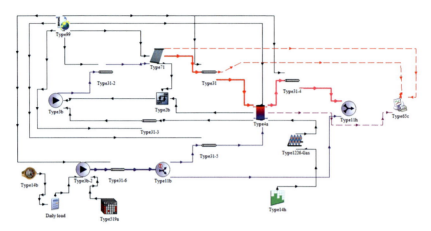

Fig. 7 TRNSYS simulation model.

heat exchanging device [45]. The fouling procedure can be calculated using Eq. (1).

$$\frac{dm_f}{dt} = \dot{m}_d - \dot{m}_r \qquad (1)$$

where dm_f, $\dot{m}_r$ and $\dot{m}_d$ are net deposition, removal, and deposition rate, respectively. Also, the deposition and removal rates act separately and combine into a net deposition rate.

Among the numerous mechanisms of fouling, crystallization fouling has been reported to be one of the primary causes of downtime in industries. Crystallization fouling is significant enough that it affects the GDP of developed countries. In an experimental examination, the supersaturated cold-water solutions of mineral salts were made by several researchers to quicken the accumulation on the heat exchanger surface to study fouling formation and mitigation. A $CaCO_3$ hard water solution was made by dissolving sodium bicarbonate and calcium chloride powders in distilled water. The formation of $CaCO_3$ by reaction are exhibited in Eq. (2):

$$CaCl_2 + NaHCO_3 \rightarrow CaCO_3 + 2NaCl + H_2O + CO_2 \qquad (2)$$

Likewise, a $CaSO_4$ solution was prepared through dissolution of calcium nitrate tetrahydrate ($Ca(NO_3)_2 \cdot 4\,H_2O$) and sodium sulfate

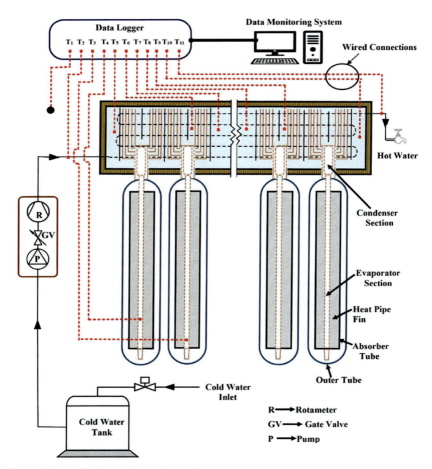

Fig. 8 Schematic view of experimental setup.

(Na_2SO_4) in stoichiometric ratios to prepare the solution of $CaSO_4$ as presented in Eq. (3).

$$Ca(NO_3)_2 \cdot 4H_2O + Na_2SO_4 \rightarrow CaSO_4 \cdot 2H_2O + 2NaNO_3 + 2H_2O \qquad (3)$$

In an experimental investigation, the supersaturated hard water of $CaCO_3$ and $CaSO_4 \cdot H_2O$ were prepared through the above reaction with very low solubility mineral salts. Mathematically, the rate of deposit growth or fouling factor (R_f) can be expressed as the difference between the accumulation and removal rates as demonstrated in Eq. (4).

$$R_f = \phi_d - \phi_r \qquad (4)$$

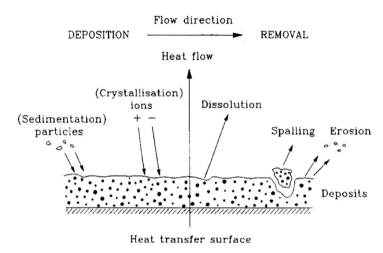

Fig. 9 Fouling mechanisms on the surface of a heat exchanger [46].

where ϕ_d is the deposition rate, ϕ_r is the removal rate and R_f is the fouling factor. The fouling factor, deposition rate, and removal rate can be expressed in the units of thermal resistance as $m^2.K/W$ or in the units of the rate of thickness change as m/s or units of mass change as kg/m^2s.

Various nanofluids can be utilized to retard scaling [47]. Teng et al. [48], found that nanofluids can function as (a) threshold agents, (b) crystal transformers, and (c) segregating agents. Some of the general additives used in water are polyphosphates and polyphosphonates (threshold agents), ethylenediaminetetraacetic acid (EDTA), and Poly carboxylic acids and derivatives (threshold agent and segregating). Segregating substances, like EDTA interact with fouling ions including Ca_2^+, Mg_2^+, and Cu_2^+ as substitutes for Na^+ to create robust complexes that inhibit and/or eliminate scaling. They are utilized efficiently for antifouling in the treatment of boiler feed water [49].

Troup and Richardson [50] reported that crystal altering substances, such as Poly carboxylic acids, alter the crystal structure and prevent the growth of bigger crystals. The altered crystals did not accumulate on the surface of a heat exchanger but continuously dispersed in the cold solution. If their concentration rises beyond a specific value, particulate scaling may occur. Particulate fouling is avoided or minimized by employing crystal alteration substances with dispersants. Though the water-based nanofluid may not inhibit the crystallization fouling completely, the final crystal deposits are discrete from the those developed in the absence of additives. The layers formed due to mineral fouling on the heat exchanger surface were weakened because of the presence of additives and can effortlessly be removed.

3.1 Impact of multiwall carbon nanotubes (MWCNT) based nanofluid on retardation of fouling and enhancement of heat transfer

Teng et al. [47,50] performed an investigation to mitigate fouling and enhance heat transfer using DPTA and EDTA treated multiwall carbon nanotubes (MWCNTs). They found encouraging results as represented in Figs. 10 and 11. DPTA and EDTA treated MWCNTs decreased the $CaCO_3$ deposition. The increase of EDTA-MWCNT concentrations in the base fluid raises the induction period because of the enhancement of additives concentration in base fluid Ca^{+2} adsorption by additives in nanofluids. Furthermore, the thermal conductivity of the nanofluid compared to water also increased which further enhanced heat transfer. This enhancement in heat transfer was due to the Brownian motion of MWCNTs and water molecules that formed nanolayer surfaces.

FE-SEM pictures of the heat exchanger surfaces at a magnification of 500X are provided in Fig. 12 to illustrate the deposition surface morphology with and without EDTA and MWCNT-EDTA nanoparticles. Sharp, pointed, and needle-like crystals developed with no additive addition; when no additives were added, bigger crystals were developed. With the additive addition, the deposits were weaker, smoother, and the size of the crystal was reduced. EDTA-MWCNT affected the heat exchanger surface with corrosion and weight loss while DPTA treated MWCNT

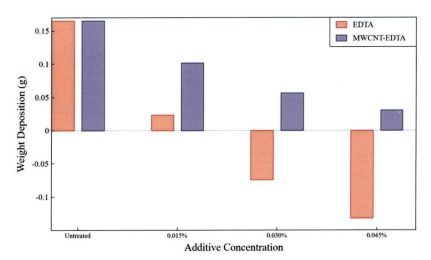

Fig. 10 $CaCO_3$ mass deposition on a SS316L heat exchanger surface with a water base fluid, EDTA based nanofluid, and EDTA-MWCNT based nanofluid [51].

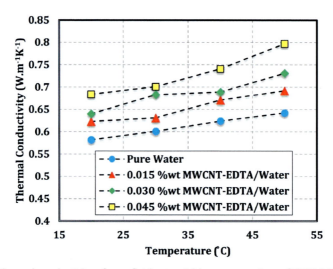

Fig. 11 Thermal conductivity of nanofluids at variable concentrations of EDTA-MWCNT [51].

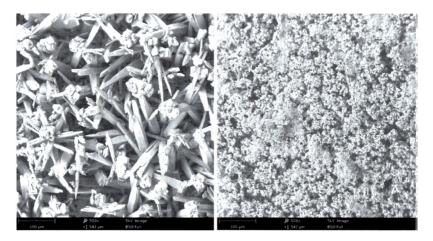

Fig. 12 FESEM images: (A) non treated with EDTA -MWCNT, (B) treated with EDTA -MWCNT [51].

demonstrated zero corrosion in comparison to EDTA treated MWCNT and provided better anti-scaling characteristics.

Shaikh et al. [49], performed a study on mitigation of $CaCO_3$ fouling and enhancement of heat transfer using clove functionalized MWCNT as nanofluids and obtained inspiring outcomes. With low concentrations of additives in water-based nanofluids, a longer induction period of fouling was achieved without any corrosion. Heat transfer with the heat exchanger increased due to greater thermal conductivity of MWCNT in comparison to the base liquid.

3.2 Impact of graphene nanoplatelet-based nanofluid (GNP) on fouling retardation and heat transfer enhancement

Shaikh et al. [52] experimentally examined transfer improvement and retardation of $CaCO_3$ mineral fouling using water-based clove functionalized graphene nanoplatelets (C-GNP) and found good results at low concentrations of C-GNP as illustrated in Figs. 13 and 14. C-GNP reduces the deposition of $CaCO_3$ on surfaces of the heat exchanger. With the enhancement of C-GNP concentrations in the base fluid, adsorption of Ca^+ by water-based nanofluids increased and a reduction of $CaCO_3$ occurred. Furthermore, with increased C-GNP concentration, the thermal conductivity of the nanofluid also increased which enhanced the heat transfer of heat exchanging device. The heat transfer displayed in Fig. 12 shows that

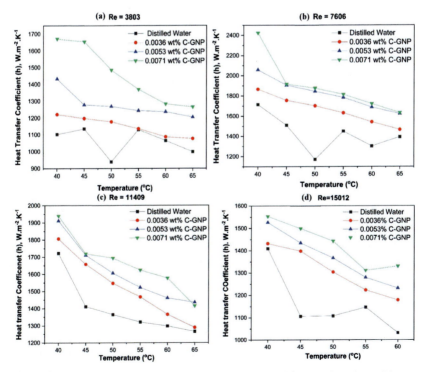

Fig. 13 Heat transfer coefficients vs temperature at Reynolds numbers 3803 (a), 7606 (b) 11409 (c) and 15010 (d).

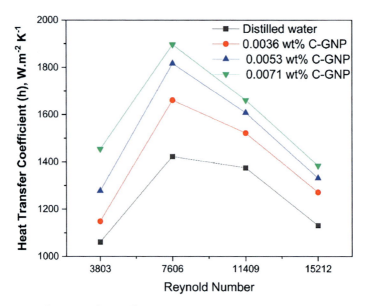

Fig. 14 Mean heat transfer coefficient variations with Reynold number.

with increased C-GNP concentration at all temperatures and Reynolds numbers, the heat transfer coefficient (h) increases. The mean heat transfer coefficient (h) at various Reynolds numbers are presented in Fig. 13; the figure shows that initial heat transfer coefficient increased with Reynolds number untilRe = 7606 and then starts to decline untilRe = 15212.

Oon et al. [53] experimentally studied mitigation of $CaCO_3$ and enhancement of the overall heat transfer coefficient with biofunctionalized (gallic acid functionalization) GNP and noticed good outcomes. The outcome of that study reveals that water-based biofunctionalized GNP reduced $CaCO_3$ deposition on surfaces of the heat exchanging device. There was a sevenfold improvement compared to the heat transfer coefficient of water.

4. Applications of nanofluids in annular flow heat exchangers

Many investigations have been conducted to enhance the thermal performance in variety of heat exchangers, few of the recent studies are done by Malazi et al. [54] numerically developed a model for spiral fin-and-tube heat exchangers (SPF HEX) for predicting heat transfer characteristics considering operation condition and Azeddine et al. [55] numerically examined thermal and dynamic properties in rectangular counter flow heat exchanger using Al_2O_3 nanofluid. Saidani et al. [56] numerically studied horizontal coaxial annular space with magneto double diffusive density driven convection through Al_2O_3 nanofluid. Olsen et al. [57] simulated the gas flow through horizontal pipe and found that for low-to-moderate Reynolds numbers, the combined effects on heat transfer and pressure loss lead to an improved performance with the obstruction and variation in properties found to increase with laminar and decrease with turbulent Nusselt number. Zhang et al. [58] investigated heat transfer in curved micro channel through Cu nanofluids and by varying curvatures. Studies have shown that annular flow with swirling fluid and recirculation can lead to improved heat transfer within an annular domain. Additionally, the use of angled fins inducing swirling flow within an annular passage has been found advantageous for heat transfer performance, especially when compared to other designs like twisted tape configurations. Overall, the unique characteristics of annular flow, such as swirling fluid dynamics and recirculation patterns, make it a favorable choice for optimizing heat transfer in various applications. Hence the annular flow regime with nanofluids is focused in this chapter. Nanofluids are mainly utilized in annular flow heat exchangers for enhanced heat transfer. It is also important to note that pressure drop penalties can occur; enhanced thermal conductivity and reduced specific heat in nanofluids improve heat transfer, however, increase density and viscosity lead to higher pressure drop. The Thermal Performance Factor (TPF) is frequently used to evaluate how well nanofluids perform as a working fluid under certain fluid flow conditions. Studies have shown that TPF values can vary based on different flow regimes and configurations. For instance, in experimental analyses of double pipe heat exchangers, the TPF was found to be less than unity for most cases, with a maximum achieved TPF close to 1.08. This indicates that the heat transfer enhancement techniques employed in such systems were moderately effective in improving overall performance. Additionally, numerical studies have demonstrated the use of TPF to assess heat transfer through various geometries like horizontal wavy surfaces, highlighting its utility in evaluating heat transfer efficiency across different designs. Overall, the Thermal Performance Factor serves as a valuable tool for quantifying the effectiveness of heat transfer enhancement methods by considering both heat transfer improvement and pressure drop implications. The enhancement techniques aim to minimize the rise in pressure drop while maximizing the improvement in heat transfer. A schematic diagram of an annular flow heat exchanger is provided in Fig. 15.

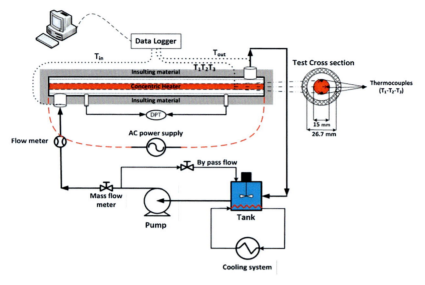

Fig. 15 Annular flow heat exchanger setup [59].

4.1 Nanofluids behavior in an annular flow heat exchanger

Both metallic and non-metallic nanofluids can be used in annular flow heat exchangers, depending on the specific requirements; M. Nasiri et al. [60] performed experiments on annular flow heat exchangers with Al_2O_3/H_2O and TiO_2/H_2O. For both nanofluids, the heat transfer improves as the concentration rises. According to the findings of this investigation, the heat transfer enhancement is almost identical when employing the same concentration of either Al_2O_3/H_2O or TiO_2/H_2O (Fig. 16).

Green and novel synthesized carbon nanotube nanofluids with covalent functionalization have been studied by Maryam Hosseini et al. both experimentally [61] and numerically [59], as shown in Fig. 17. To remove the usage of harmful acids in typical carbon nanoparticle functionalization processes, this study introduced a simple and environmentally friendly procedure for synthesizing highly dispersed C-MWCNTs. The outcomes demonstrated that there is a notable increase in the convective heat transfer coefficient and Nusselt number of 35.89% and 20.15%, respectively. The substantial improvement occurs at a Reynolds number of 7944, a nano additive concentration of 0.175 wt%, and a constant heat flow of 38,346 W/m^2. The friction factor of C-MWCNT-DI water nanofluids can increase by up to 3% related to the base fluid when the nanoparticle concentration is 0.175 wt%. The performance index of the C-MWCNT-DI

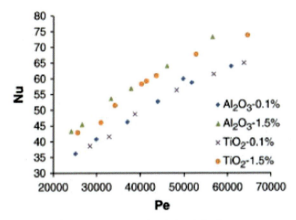

Fig. 16 Nu comparison for Al$_2$O$_3$/H$_2$O and TiO$_2$/H$_2$O [60].

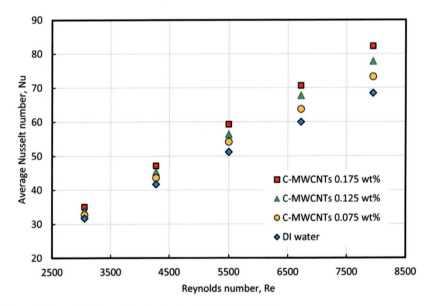

Fig. 17 Nu with Reynolds number in C-MWCNT-DI [61].

water nanofluids is consistently above 1, regardless of the nonadditive concentration studied. This suggests that nanofluids have the capability to be effective working fluids in heat transfer systems, offering improved thermal performance and energy efficiency.

While often water is utilized as a base fluid, other liquids can be used as well. Cobanoglu et al. [62] used water-ethylene glycol (WEG) mixture-based Fe$_3$O$_4$ nanofluids in the natural circulation of annular flow heat exchangers. Apart from the effects of heat transfer and pressure drop, exergy destruction was also incorporated into this study (Fig. 18).

Graphene nanoplatelets (GNPs) in water/ethylene glycol base fluid have been numerically simulated in the annular flow heat exchanger by Khajeh Arzani et al. [63]. An analysis of turbulent forced convection heat transfer under a constant heat flow has been conducted. This novel type of working fluid is suggested to be a highly efficient coolant for annular heat exchangers because of its superior thermal characteristics and potential for energy conservation. All samples exhibited greater thermal conductivity, viscosity, and density in comparison to the basic fluid. Conversely, the

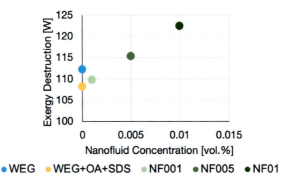

Fig. 18 Exergy destruction [62].

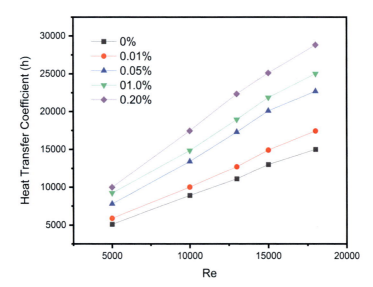

Fig. 19 Heat transfer coefficient of GNP [63].

specific heat capacity drops significantly. Minimal increase in pressure drop across numerous concentrations and suitable performance index are very advantageous features for introducing a new fluid for diverse industrial applications, including its use in annular heat exchangers (Fig. 19).

Although nanofluid has not yet been widely adopted in industrial and commercial applications, it continues to be a subject of fascination for researchers. The primary factors driving its appeal are its superior thermal conductivity properties, which offer boundless possibilities for improving heat transfer rates, thanks to a wide range of available nanoparticles. Nevertheless, the main obstacle preventing the practical application of nanofluids in industry is the problem of instability. Achieving the ideal size, shape, and concentration of nanoparticles is crucial for creating nanofluids that operate well and exhibit sufficient stability. For the nanofluid to have the best thermal capabilities, it is important to prioritize the selection of a compatible base fluid, surfactant, and the appropriate quantity of surfactant. To reduce significant decreases in pressure while giving priority to improving heat transmission, various strategies can be utilized, such as introducing holes, reducing obstruction, or manipulating the orientation of modified tubes, to achieve higher thermal performance factor values.

5. Conclusion

Heat exchangers are common in many industries and applications such as power plants, petrochemical applications, paper industry, and food industries, etc. The thermal efficiency of heat exchanger with water (i.e., lakes, rivers, and ground water) is very low because of the thermal conductivity of water and impurities present in raw water. To increase the thermal performance of heat exchangers several studies on different nanofluids (such as metal oxides, carbon, and composite based nanofluids) have been conducted. The present study focused on three aspects of nanofluids in heat exchangers. One focus is to improve the thermal performance of solar collectors. A second is to reduce the fouling on heat exchanger surfaces and enhance heat transfer. The third aspect is to use nanofluids in the annular flow heat exchangers. Studies conducted on metal oxide and its composite reveal that heat transfer performance in solar collectors and annular heat exchangers is enhanced due to water-based metal oxides, and it composites nanofluids. However, these nanofluids can be hazardous to the environment and to human health.

To resolve this environmental hazard problem, researchers have seriously taken efforts to synthetize the ecofriendly carbon based nanofluids such as MWCNT, GNP, activated carbon nanoparticles and their hybrids etc. Researchers have observed that with carbon based nanofluids not only reduced the environmental hazards but also increased the heat transfer and mitigation of fouling was noticed. Carbon based nanofluids also prolong heat exchanger life.

Acknowledgements

This study was supported financially under HORIZON-MSCA-2022-SE-01-01 awarded by European Research Executive Agency with project number 101130406 (ice-Link). The authors also greatfully acknowledge the support from grants [RMF0400-2021, ST049-2022, RK001-2022], centers (Nanotechnology and Catalysis Research Center (NANOCAT), AMMP Centre, Centre for Energy Sciences) and Department of Mechanical Engineering, Universiti Malaya to conduct this work.

References

[1] K.R.V. Subramanian, T.N. Rao, A. Balakrishnan (Eds.), Nanofluids and Their Engineering Applications, first ed., CRC Press, 2019. https://doi.org/10.1201/9780429468223.

[2] H. Mamat, M. Ramadan, Nanofluids: thermal conductivity and applications, Encyclopedia of Smart Materials, Elsevier, 2022, pp. 288–296, https://doi.org/10.1016/B978-0-12-815732-9.00141-8.

[3] J. Li, X. Zhang, B. Xu, M. Yuan, Nanofluid research and applications: a review, Int. Commun. Heat. Mass. Transf. 127 (2021) 105543, https://doi.org/10.1016/j.icheatmasstransfer.2021.105543

[4] J.-H. Ku, H.-H. Cho, J.-H. Koo, S.-G. Yoon, J.-K. Lee, Heat transfer characteristics of liquid-solid suspension flow in a horizontal pipe, KSME Int. J. 14 (2000) 1159–1167, https://doi.org/10.1007/BF03185070

[5] A.S. Ahuja, Augmentation of heat transport in laminar flow of polystyrene suspensions. I. Experiments and results, J. Appl. Phys. 46 (1975) 3408–3416, https://doi.org/10.1063/1.322107

[6] A.S. Ahuja, Augmentation of heat transport in laminar flow of polystyrene suspensions. II. Analysis of the data, J. Appl. Phys. 46 (1975) 3417–3425, https://doi.org/10.1063/1.322062

[7] S.U.S. Choi, J.A. Eastman, Enhancing Thermal Conductivity of Fluids with Nanoparticles, University of North Texas Libraries, UNT Digital Library, San Francisco, CA, 1995.

[8] J.A. Eastman, S.U.S. Choi, S. Li, L.J. Thompson, S. Lee, Enhanced thermal conductivity through the development of nanofluids, MRS Proc. 457 (1996) 3, https://doi.org/10.1557/PROC-457-3

[9] Y. Hwang, J.-K. Lee, J.-K. Lee, Y.-M. Jeong, S. Cheong, Y.-C. Ahn, et al., Production and dispersion stability of nanoparticles in nanofluids, Powder Technol. 186 (2008) 145–153, https://doi.org/10.1016/j.powtec.2007.11.020

[10] X. Wang, X. Xu, S.U.S. Choi, Thermal conductivity of nanoparticle – fluid mixture, J. Thermophys. Heat. Transf. 13 (1999) 474–480, https://doi.org/10.2514/2.6486

[11] P. Jaiswal, Y. Kumar, L. Das, V. Mishra, R. Pagar, D. Panda, et al., Nanofluids guided energy-efficient solar water heaters: recent advancements and challenges ahead, Mater. Today Commun. 37 (2023) 107059, https://doi.org/10.1016/j.mtcomm.2023.107059

[12] R. Taylor, S. Coulombe, T. Otanicar, P. Phelan, A. Gunawan, W. Lv, et al., Small particles, big impacts: a review of the diverse applications of nanofluids, J. Appl. Phys. 113 (2013) 011301, https://doi.org/10.1063/1.4754271

[13] R. Saidur, K.Y. Leong, H.A. Mohammed, A review on applications and challenges of nanofluids, Renew. Sustain. Energy Rev. 15 (2011) 1646–1668, https://doi.org/10.1016/j.rser.2010.11.035

[14] G. Huminic, A. Huminic, Application of nanofluids in heat exchangers: a review, Renew. Sustain. Energy Rev. 16 (2012) 5625–5638, https://doi.org/10.1016/j.rser.2012.05.023

[15] E.V. Timofeeva, W. Yu, D.M. France, D. Singh, J.L. Routbort, Nanofluids for heat transfer: an engineering approach, Nanoscale Res. Lett. 6 (2011) 182, https://doi.org/10.1186/1556-276X-6-182

[16] E.V. Timofeeva, Nanofluids for heat transfer – potential and engineering strategies, in: A. Ahsan (Ed.), Two Phase Flow, Phase Change and Numerical Modeling, InTech, 2011. https://doi.org/10.5772/22158.

[17] M. Awais, A.A. Bhuiyan, S. Salehin, M.M. Ehsan, B. Khan, Md.H. Rahman, Synthesis, heat transport mechanisms and thermophysical properties of nanofluids: a critical overview, Int. J. Thermofluids 10 (2021) 100086, https://doi.org/10.1016/j.ijft.2021.100086

[18] B. Mehta, D. Subhedar, H. Panchal, Z. Said, Synthesis, stability, thermophysical properties and heat transfer applications of nanofluid – a review, J. Mol. Liq. 364 (2022) 120034, https://doi.org/10.1016/j.molliq.2022.120034

[19] S.U.S. Choi, Nanofluids: from vision to reality through research, J. Heat. Transf. 131 (2009) 033106, https://doi.org/10.1115/1.3056479

[20] S. Chakraborty, P.K. Panigrahi, Stability of nanofluid: a review, Appl. Therm. Eng. 174 (2020) 115259, https://doi.org/10.1016/j.applthermaleng.2020.115259

[21] F. Yu, Y. Chen, X. Liang, J. Xu, C. Lee, Q. Liang, et al., Dispersion stability of thermal nanofluids, Prog. Nat. Sci.: Mater. Int. 27 (2017) 531–542, https://doi.org/10.1016/j.pnsc.2017.08.010

[22] S.A. Kalogirou, Solar thermal collectors and applications, Prog. Energy Combust. Sci. 30 (2004) 231–295, https://doi.org/10.1016/j.pecs.2004.02.001

[23] S.A. Kalogirou, Solar thermal energy: history, in: S. Alexopoulos, S.A. Kalogirou (Eds.), Solar Thermal Energy, Springer US, New York, NY, 2022, pp. 7–19. https://doi.org/10.1007/978-1-0716-1422-8_1106.

[24] Y. Kumar, S. Kumar, P. Verma, Role of absorber and glazing in thermal performance improvements of liquid flat plate solar collector: a review, Energy Sources Part A: Recov. Util. Environ. Eff. 45 (2023) 10802–10826, https://doi.org/10.1080/15567036.2023.2249849

[25] M. Kanoglu, Y.A. Çengel, J.M. Cimbala, Fundamentals and Applications of Renewable Energy, McGraw-Hill Education, New York, NY, 2020.

[26] A. Aissa, N.A.A. Qasem, A. Mourad, H. Laidoudi, O. Younis, K. Guedri, et al., A review of the enhancement of solar thermal collectors using nanofluids and turbulators, Appl. Therm. Eng. 220 (2023) 119663, https://doi.org/10.1016/j.applthermaleng.2022.119663

[27] M.J. Alshukri, A. Kadhim Hussein, A.A. Eidan, A.I. Alsabery, A review on applications and techniques of improving the performance of heat pipe-solar collector systems, Sol. Energy 236 (2022) 417–433, https://doi.org/10.1016/j.solener.2022.03.022

[28] F. Masood, N.B.M. Nor, P. Nallagownden, I. Elamvazuthi, R. Saidur, M.A. Alam, et al., A review of recent developments and applications of compound parabolic concentrator-based hybrid solar photovoltaic/thermal collectors, Sustainability 14 (2022) 5529, https://doi.org/10.3390/su14095529

[29] S. Sreekumar, N. Shah, J.D. Mondol, N. Hewitt, S. Chakrabarti, Broadband absorbing mono, blended and hybrid nanofluids for direct absorption solar collector: a comprehensive review, Nano Futures 6 (2022) 022002, https://doi.org/10.1088/2399-1984/ac57f7

[30] O. Ayadi, O. Al–Oran, M. Hamdan, T. Salameh, A.A. Hasan, A. Juaidi, et al., Utilization of mono and hybrid nanofluids in solar thermal collectors, Renewable Energy Production and Distribution, Elsevier, 2022, pp. 3–44. https://doi.org/10.1016/B978-0-323-91892-3.00008-X.

[31] Q. Xiong, A. Hajjar, B. Alshuraiaan, M. Izadi, S. Altnji, S.A. Shehzad, State-of-the-art review of nanofluids in solar collectors: a review based on the type of the dispersed nanoparticles, J. Clean. Prod. 310 (2021) 127528, https://doi.org/10.1016/j.jclepro.2021.127528

[32] H. Younes, M. Mao, S.M. Sohel Murshed, D. Lou, H. Hong, G.P. Peterson, Nanofluids: key parameters to enhance thermal conductivity and its applications, Appl. Therm. Eng. 207 (2022) 118202, https://doi.org/10.1016/j.applthermaleng.2022.118202

[33] H. Wang, X. Li, B. Luo, K. Wei, G. Zeng, The MXene/water nanofluids with high stability and photo-thermal conversion for direct absorption solar collectors: a comparative study, Energy 227 (2021) 120483, https://doi.org/10.1016/j.energy.2021.120483

[34] S.M.S. Hosseini, M.S. Dehaj, The comparison of colloidal, optical, and solar collection characteristics between Fe_2O_3 and Fe_3O_4 nanofluids operated in an evacuated tubular volumetric absorption solar collector, J. Taiwan. Inst. Chem. Eng. 135 (2022) 104381, https://doi.org/10.1016/j.jtice.2022.104381

[35] O.A. Hussein, K. Habib, A.S. Muhsan, R. Saidur, O.A. Alawi, T.K. Ibrahim, Thermal performance enhancement of a flat plate solar collector using hybrid nanofluid, Sol. Energy 204 (2020) 208–222, https://doi.org/10.1016/j.solener.2020.04.034

[36] A. Moradi, M. Zareh, M. Afrand, M. Khayat, Effects of temperature and volume concentration on thermal conductivity of TiO2-MWCNTs (70-30)/EG-water hybrid nano-fluid, Powder Technol. 362 (2020) 578–585, https://doi.org/10.1016/j.powtec.2019.10.008

[37] X. Li, G. Zeng, X. Lei, The stability, optical properties and solar-thermal conversion performance of SiC-MWCNTs hybrid nanofluids for the direct absorption solar collector (DASC) application, Sol. Energy Mater. Sol. Cell 206 (2020) 110323, https://doi.org/10.1016/j.solmat.2019.110323

[38] R. Liang, J. Zhang, L. Zhao, L. Ma, Research on the universal model of filled-type evacuated tube with U-tube in uniform boundary condition, Appl. Therm. Eng. 63 (2014) 362–369, https://doi.org/10.1016/j.applthermaleng.2013.11.020

[39] D.N. Korres, C. Tzivanidis, I.P. Koronaki, M.T. Nitsas, Experimental, numerical and analytical investigation of a U-type evacuated tube collectors' array, Renew. Energy 135 (2019) 218–231, https://doi.org/10.1016/j.renene.2018.12.003

[40] M.A. Ersöz, Effects of different working fluid use on the energy and exergy performance for evacuated tube solar collector with thermosyphon heat pipe, Renew. Energy 96 (2016) 244–256, https://doi.org/10.1016/j.renene.2016.04.058

[41] G. Sadeghi, M. Najafzadeh, M. Ameri, Thermal characteristics of evacuated tube solar collectors with coil inside: an experimental study and evolutionary algorithms, Renew. Energy 151 (2020) 575–588, https://doi.org/10.1016/j.renene.2019.11.050

[42] A. Mehmood, A. Waqas, Z. Said, S.M.A. Rahman, M. Akram, Performance evaluation of solar water heating system with heat pipe evacuated tubes provided with natural gas backup, Energy Rep. 5 (2019) 1432–1444, https://doi.org/10.1016/j.egyr.2019.10.002

[43] P. Selvakumar, P. Somasundaram, P. Thangavel, Performance study on evacuated tube solar collector using therminol D-12 as heat transfer fluid coupled with parabolic trough, Energy Convers. Manag. 85 (2014) 505–510, https://doi.org/10.1016/j.enconman.2014.05.069

[44] K. Chopra, V.V. Tyagi, A.K. Pathak, A.K. Pandey, A. Sari, Experimental performance evaluation of a novel designed phase change material integrated manifold heat pipe evacuated tube solar collector system, Energy Convers. Manag. 198 (2019) 111896, https://doi.org/10.1016/j.enconman.2019.111896

[45] S.N. Kazi, Particulate matter: interfacial properties, fouling, and its mitigation, Water-Formed Deposits, Elsevier, 2022, pp. 97–140, https://doi.org/10.1016/B978-0-12-822896-8.00012-1

[46] S.N. Kazi, Fouling and fouling mitigation on heat exchanger surfaces, in: J. Mitrovic (Ed.), Heat Exchangers – Basics Design Applications, InTech, 2012. https://doi.org/10.5772/32990.

[47] K. Shaikh, K.M.S. Newaz, M.N.M. Zubir, K.H. Wong, W.A. Khan, S. Abdullah, et al., A review of recent advancements in the crystallization fouling of heat exchangers, J. Therm. Anal. Calorim. (2023), https://doi.org/10.1007/s10973-023-12544-z

[48] K.H. Teng, A. Amiri, S.N. Kazi, M.A. Bakar, B.T. Chew, A. Al-Shamma'a, et al., Retardation of heat exchanger surfaces mineral fouling by water-based diethylene-triamine pentaacetate-treated CNT nanofluids, Appl. Therm. Eng. 110 (2017) 495–503, https://doi.org/10.1016/j.applthermaleng.2016.08.181

[49] K. Shaikh, S.N. Kazi, M.N.M. Zubir, K.H. Wong, W.A. Khan, S. Abdullah, et al., Retardation of $CaCO_3$ fouling on heat exchanger surface using water-based cloves-functionalized multiwall carbon nanotubes (C-MWCNT) nanofluids, J. Therm. Anal. Calorim. (2023), https://doi.org/10.1007/s10973-023-12551-0

[50] D.H. Troup, J.A. Richardson, Scale nucleation on A heat transfer surface and its prevention, Chem. Eng. Commun. 2 (1978) 167–180, https://doi.org/10.1080/00986447808960458

[51] K.H. Teng, A. Amiri, S.N. Kazi, M.A. Bakar, B.T. Chew, Fouling mitigation on heat exchanger surfaces by EDTA-treated MWCNT-based water nanofluids, J. Taiwan. Inst. Chem. Eng. 60 (2016) 445–452, https://doi.org/10.1016/j.jtice.2015.11.006

[52] K. Shaikh, S.N. Kazi, M.N.M. Zubir, K.H. Wong, W.A. Khan, S. Abdullah, et al., Improvement of heat transfer and fouling retardation performance of heat exchanger using green functionalized graphene nanoplatelets additives, J. Taiwan. Inst. Chem. Eng. 153 (2023) 105227, https://doi.org/10.1016/j.jtice.2023.105227

[53] C.S. Oon, S.N. Kazi, N. Zubir, I.A. Badruddin, S. Kamangar, C.Y. Heah, et al., Fouling and fouling mitigation of mineral salt using bio-based functionalized graphene nano-plates, J. Therm. Anal. Calorim. 146 (2021) 265–275, https://doi.org/10.1007/s10973-020-09940-0

[54] M.T. Malazi, S. Apak, C. Sezer, H.C. Başoğlu, A.S. Dalkılıç, Investigation of the thermal performance of a spiral fin-and-tube heat exchanger by numerical methods, Numer. Heat. Transf. Part B: Fundamentals (2024) 1–21, https://doi.org/10.1080/10407790.2024.2316196

[55] S. Azeddine, D. Saadaoui, I. Rahmoune, Computational investigation on the energy efficiency of heat exchangers using Al_2O_3 nanofluids, Numer. Heat. Transf. Part A: Appl. (2024) 1–21, https://doi.org/10.1080/10407782.2024.2329313

[56] L. Saidani, T. Tayebi, M. Djezzar, E.H. Malekshah, Magneto-double-diffusive natural convection and irreversibility analysis of a nanofluid flowing in an annular concentric space, Numer. Heat. Transf. Part A: Appl. (2023) 1–23, https://doi.org/10.1080/10407782.2023.2272787

[57] L. Olsen, S. Bhattacharyya, L. Cheng, W. Minkowycz, J. Abraham, Heat transfer enhancement for internal flows with a centrally located circular obstruction and the impact of buoyancy, Heat. Transf. Eng. 43 (2022) 1789–1805, https://doi.org/10.1080/01457632.2021.2016129

[58] X. Zhang, Y. Zhang, C. Yang, J. Liu, B. Liu, Z. Zhou, Study of flow and heat transfer performance of nanofluids in curved microchannels with different curvatures, Numer. Heat. Transf. Part A: Appl. (2024) 1–22, https://doi.org/10.1080/10407782.2024.2321520

[59] M. Hosseini, A.H. Abdelrazek, R. Sadri, A.R. Mallah, S.N. Kazi, B.T. Chew, et al., Numerical study of turbulent heat transfer of nanofluids containing eco-friendly treated carbon nanotubes through a concentric annular heat exchanger, Int. J. Heat. Mass. Transf. 127 (2018) 403–412, https://doi.org/10.1016/j.ijheatmasstransfer.2018.08.040

[60] M. Nasiri, Etemad SGh, R. Bagheri, Experimental heat transfer of nanofluid through an annular duct, Int. Commun. Heat. Mass. Transf. 38 (2011) 958–963, https://doi.org/10.1016/j.icheatmasstransfer.2011.04.011

[61] M. Hosseini, R. Sadri, S.N. Kazi, S. Bagheri, N. Zubir, C. Bee Teng, et al., Experimental study on heat transfer and thermo-physical properties of covalently functionalized carbon nanotubes nanofluids in an annular heat exchanger: a green and novel synthesis, Energy Fuels 31 (2017) 5635–5644, https://doi.org/10.1021/acs.energyfuels.6b02928

[62] N. Çobanoğlu, A. Banisharif, P. Estellé, Z.H. Karadeniz, The developing flow characteristics of water – ethylene glycol mixture based Fe_3O_4 nanofluids in eccentric annular ducts in low temperature applications, Int. J. Thermofluids 14 (2022) 100149, https://doi.org/10.1016/j.ijft.2022.100149

[63] H. Khajeh Arzani, A. Amiri, H.K. Arzani, S.B. Rozali, S.N. Kazi, A. Badarudin, Toward improved heat transfer performance of annular heat exchangers with water/ethylene glycol-based nanofluids containing graphene nanoplatelets, J. Therm. Anal. Calorim. 126 (2016) 1427–1436, https://doi.org/10.1007/s10973-016-5663-8

CHAPTER FOUR

Programmable micro- and nano-engineered liquid metals in thermal engineering applications

Rahul Agarwal[a], Saleh S. Baakeem[b], and A.A. Mohamad[a],*

[a]Department of Mechanical and Manufacturing Engineering, University of Calgary, Calgary, AB, Canada
[b]Department of Chemical and Petroleum Engineering, University of Calgary, Calgary, AB, Canada
*Corresponding author. e-mail address: mohamad@ucalgary.ca

Contents

1. Introduction	130
2. Morphological forms	132
2.1 Bulk (droplet) form	132
2.2 Particle form	133
2.3 Liquid metal marbles	133
3. Processing techniques	134
3.1 Drop-on-demand	134
3.2 Molding	134
3.3 Microfluidics	135
3.4 Sonication	136
3.5 Shearing	136
3.6 Liquid metal marbles	137
4. Actuation principles and associated characteristics of liquid metals	137
4.1 Mechanical actuation	140
4.2 Electrical actuation	141
4.3 Electrocapillarity	141
4.4 Magnetic actuation	148
4.5 Optical actuation	149
4.6 Acoustic actuation	150
4.7 Thermal actuation	151
4.8 Chemical actuation	153
4.9 Self-propelling actuation	154
4.10 Multi-stimuli actuation	155
5. Applications of liquid metals in thermal engineering	156
5.1 Phototherapeutic	156
5.2 Thermal switches	158
5.3 Photothermal actuators	162
5.4 Energy harvesting	163
5.5 Microfluidic thermal management	165
5.6 Thermal interface materials	166

Advances in Heat Transfer, Volume 57
ISSN 0065-2717, https://doi.org/10.1016/bs.aiht.2023.12.001
Copyright © 2024 Elsevier Inc. All rights are reserved, including those for text and data mining, AI training, and similar technologies.

5.7 Composite materials	168
5.8 Microencapsulated phase change materials	170
6. Our perspective and future directions	173
References	174

Abstract

In recent years, liquid metals (LMs) have garnered increasing attention as newly emerging functional materials in thermal management. These metals exhibit fascinating properties such as high surface tension, high electrical and thermal conductivity, phase transition phenomenon, low viscosity and vapor pressure, non-toxicity, and biocompatibility. Notably, there have been significant advancements across different categories. These include high-performance LM-based convection cooling technology, low melting point phase change materials, thermal interface materials, and energy harvesters/heat sinks. LMs, with their remarkable heat extraction and transport capabilities compared to conventional coolants, are poised to overcome the limitations currently faced by existing thermal strategies. This chapter provides a comprehensive review of the LMs at all researcher levels, from beginners to experts. It delves into various aspects of the LMs, including morphology, advances, processes, applications, etc. Furthermore, we offer an in-depth exploration of the progress in the sciences and technologies enabled by LMs in thermal management. Also, the chapter provides an account of the various challenges in this direction, such as corrosion and compatibility, chemical stability, thermal conductivity, low-temperature operations, and wettability. These challenges are crucial from a scientific perspective and the prospects of this exciting field. Further, these insights are pivotal in the development of LM-based devices.

1. Introduction

Semiconductor chips can generate high heat flux resulting in elevated junction temperatures, making heat dissipation paramount. This rise in the temperature of the internal component temperatures reduces the performance and may even lead to accidents. Reports indicate heat–related issues contribute to over 50% of integrated circuit (IC) failures. It has been reported that for every 10 °C decrease in temperature, the failure rate of electronic components may be halved, and component reliability increases with decreasing operating temperatures [1]. Moreover, the significant strides in miniaturization, integration, and power density have led to a substantial increase in heat generation.

With this rapid advancement of microelectronic devices, the "more–than–Moore" strategy has emerged as a pivotal trend in surmounting previous limitations of IC and spurring innovation in traditional cooling methods. These traditional cooling methods include forced air cooling [2], forced water cooling [3], phase change material (PCM) cooling [4], and thermal grease [5]. However, due to the thermophysical properties of these heat transfer mediums, their cooling capacity struggles to meet the rapidly increasing heat dissipation requirements. Therefore, developing new cooling media is at the core of innovation in cooling technology [6,7].

Recently, there has been a growing focus on new advanced thermal functional materials such as liquid metals (LMs). Generally, a substance composed of a single metallic element or alloy with a low melting point and a liquid phase at room temperature is classified as a "liquid metal". These LMs with low melting points have garnered increasing popularity due to their exceptional properties, such as fluidity, flexibility, wettability, and non-toxicity. Moreover, they have outstanding thermal characteristics, including high thermal conductivity and a high latent heat per unit volume, making them a favorite in emerging thermal management approaches. These qualities make the LMs a promising material for various applications, including 3D printing [8], soft robots [9,10], flexible and wearable devices [11], biomedicine [12], and solar energy systems [13]. Fig. 1 summarizes a timeline of the development history and application field of LMs. The mercury (Hg) is considered an LM, which has a melting point of

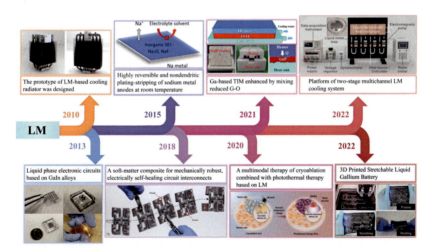

Fig. 1 Timeline of the liquid metal development and application fields. *Reproduced from Ref. S. Wang, X. Zhao, J. Luo, L. Zhuang, D. Zou, Liquid metal (LM) and its composites in thermal management, Compos. Part A Appl. Sci. Manuf. 163 (2022) 107216, https://doi.org/10.1016/j.compositesa.2022.107216, with permission obtained from Elsevier [14].*

−38.86 °C. Though used in the electronics and medical industries, it is characterized by its extreme toxicity, resulting in stringent use regulations [15]. In contrast to traditional metals like mercury, gallium (Ga) and its alloys are non-toxic metallic materials. Another advantage of Ga over other LMs is its high thermal conductivity, which reaches approximately 33.68 W·m^{-1}·K^{-1} [16]. This opens possibilities for future applications, particularly in thermal management. The current work focuses on gallium-based LMs (GaLMs). From this point forward, the terms LMs and GaLMs are used interchangeably in the text.

In this chapter, Section 2 reviews the three morphological aspects of LMs, (drops, particles, and marbles). We then provide details of the recent advances in the preparation methods and processes of each form of LMs in Section 3. An in-depth exploration of recent advances in actuating LMs is given in Section 4 and the applications of LMs in thermal engineering are provided in Section 5. Finally, we summarize the challenges and prospects of LMs in the thermal management system in Section 6.

2. Morphological forms

There are three typical morphologies of gallium-based LM, as shown in Fig. 2. The description and details will be discussed in the following section.

2.1 Bulk (droplet) form

A bulk form of LM is defined as a single-continuous volume or stream that possesses metallic properties (electrical, thermal, optical, and chemical) while exhibiting fluidic attributes such as flowability and wetting behavior.

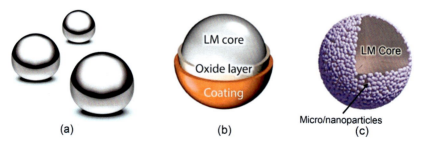

Fig. 2 The morphological forms of liquid metals (LMs) are addressed in the present work for utilization in thermal engineering. (A) LM drop, (B) LM drop with oxidized shell, and (C) LM drop coated with micro/nanoparticles.

The significance of the interfacial force relative to the body force depends on the length scale of the system. If the characteristic length of an LM volume exceeds its capillary length, the gravitational force acting on the metal can surpass the interfacial force, leading to lateral flow. A pictorial depiction of a bulk volume of LM is shown in Fig. 2A.

The bulk form of LM has unique properties, such as low viscosity, high density, high surface tension, and metallic conductivity. It is used as a conductive fluid even at room temperature, establishing electrical contacts on surfaces [17]. It creates soft and stretchable electronic components encased in an elastomer, such as wires, antennas, and circuits [18,19]. When used in an electrolyte, the electrically conductive LM interface can generate Marangoni force (tension gradient) [20]. In terms of surface tension, the LM has high surface tension that is electrically tuned, which allows for the exploration of fluidic phenomena electrically, which is not achievable in common fluids like water without changing the length scales or velocities of fluid streams [21,22]. Since LM has high metallic thermal conductivity, it is used in heat transfer, thermal switches, and reconfigurable circuits [23].

2.2 Particle form

The second morphology form of LM is particles with an oxidized nano-layer, shown in Fig. 2B. These particles have a high surface area advantageous for catalyzing or initiating reactions, plasmonic optical characteristics at small length scales, and unique phase behavior that improves supercooling at small particle sizes. Furthermore, they can be easily transformed into colloidal droplets due to their low viscosity. They can provide a metallic alternative to conventional polymeric or oxide colloids while preserving their soft mechanical properties. The particles tend to maintain their shape once generated and do not merge unless the native oxide is removed from the particles.

2.3 Liquid metal marbles

The physical coating of ferromagnetic, semiconductors, and insulating materials, whether in the form of nano or microparticles, onto the surfaces of LMs to create what is popularly referred to in the literature as "liquid metal marbles" [24–30] (Fig. 2C). This term stems from "liquid marble," originally coined almost a decade ago to describe droplets of aqueous fluid enclosed within hydrophobic particles [31].

3. Processing techniques

Thanks to their low viscosity, LMs can readily be fragmented into colloidal droplets through various techniques. Recent comprehensive reviews have extensively delved into utilizing such droplets, encompassing their roles in nanotechnology, energy harvesting, catalysis, and other domains [16]. The use of these droplets presents a compelling prospect, primarily owing to their high surface area, which is advantageous for catalytic or initiatory reactions [32]. Additionally, they offer a metallic alternative to conventional polymeric or oxide colloids while preserving favorable soft mechanical properties. The following discussion will explore the methods for producing LM drops spanning a broad range of sizes.

3.1 Drop-on-demand

The drop-on-demand (DoD) method leverages the fluidic properties of LMs by gently extruding them from a syringe or nozzle. This technique is highly regarded for its simplicity in fabricating LM droplets (LMDs) [16] and is frequently employed for crafting intricate 3D structures. Moreover, the DoD method shows a low polydispersity, typically falling within the range of 1.7%–3.1% [33]. Ladd et al. showcased a direct-writing approach capable of printing 3D structures at room temperature [34]. In this study, the researchers utilized EGaIn, which rapidly forms a gallium oxide when expelled from the syringe. This oxide layer provides essential mechanical support, ensuring the structural integrity of the 3D-printed object. Consequently, DoD enables the rapid production of self-supporting structures, wires, microfluidic channels, and, notably, LM drops [32]. Ensuring a consistent and controlled production of LMDs becomes more demanding, and the size of LMDs manufactured using DoD is constrained from approximately 70 µm to several millimeters [32].

3.2 Molding

Molding represents another method for producing LMDs, typically spanning a range from tens of micrometers to a few millimeters in size [35]. This technique entails prefabricated molds capable of shaping the LM into droplets as small as 1 µm [32]. Drawing inspiration from microfluidic fabrication processes, molds with varying reservoir sizes can be crafted. The LM alloy is subsequently forced into these patterned voids within the mold, resulting in droplets whose measures align with the volume of the reservoirs. This process is visually illustrated in Fig. 3A and B, showcasing the

Micro- and nano-engineered liquid metals 135

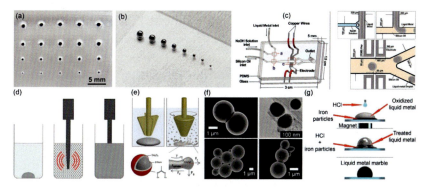

Fig. 3 Processing techniques of liquid metal droplets. (A) and (B) EGaIn droplets fabricated by pressing into a PDMS mold before and after release [35]. (C) Schematic view of a polydimethylsiloxane (PDMS)/glass microfluidic chip [36]. (D) Schematic illustration of the production of LM nanodroplets using ultrasonication. (E) Schematic illustration of the process leading to the transformation of liquid metal into micro- and nanodroplets subjected to applied shearing. (F) SEM and TEM images of shear-generated liquid metal micro and nanodroplets [39]. (G) Conceptual schematics of chemical coating with iron particles and HCl solution onto the liquid metal surface. *(D) Reproduced from Ref. K. Akyildiz, J.-H. Kim, J.-H. So, H.-J. Koo, Recent progress on micro- and nanoparticles of gallium-based liquid metals: from preparation to applications, J. Ind. Eng. Chem. 116 (2022) 120–141, https://doi.org/10.1016/j.jiec.2022.09.046, with permission obtained from Elsevier [37]. (E, F) Reproduced from I.D. Tevis, L.B. Newcomb, M. Thuo, Synthesis of liquid core–shell particles and solid patchy multicomponent particles by shearing liquids into complex particles (SLICE), Langmuir 30 (47) (2014) 14308–14313, https://doi.org/10.1021/la5035118, with permission obtained from ACS Publications.*

generation of droplets ranging from hundreds of micrometers to several millimeters. Another notable advantage of this method is the attainment of a perfectly uniform LMD size distribution [35]. Nevertheless, this introduces the challenge of effectively detaching the metal from the mold due to its inherent adhesion to the metal coating. Despite these considerations, molding remains a straightforward technique offering precise control over droplet size and size distribution.

3.3 Microfluidics

Microfluidic generation is a precise approach for producing LMDs with a low polydispersity, typically falling within 1%–3% [32]. It involves the application of a fluid that is immiscible with LM to disintegrate a stream of LM into smaller LMDs [40]. Presently, there are three distinct microfluidic methods for LMD generation: the co-flowing device, the T-shaped junction, and the flow-focusing apparatus [41]. The co-flowing device comprises two concentric microfluidic channels. LM flows in a dispersed fashion within the inner channel, while the outer channel carries a continuous phase of carrier fluid [38]. Tian et al. [36]. proposed a microfluidic system for generating and sorting Galinstan droplets, shown in Fig. 3C. To overcome LM's surface tension, it is necessary to employ viscous fluids as the carrier fluid. Gol et al. [42]. demonstrated a microfluidic system capable of producing pristine, unoxidized Galinstan droplets by harnessing interfacial hydrodynamic lift and surface tension gradient forces. Consequently, the smallest achievable diameter for LMDs using this method is limited to tens of micrometers [43].

3.4 Sonication

Sonication is a highly effective technique for generating small LMDs, typically ranging from tens of nanometers to a few micrometers in size [44]. Ultrasonication is a process harnessing sound waves to induce a vapor cavity, which, in turn, gives rise to the formation of LMDs within an aqueous medium [45], as depicted in Fig. 3D. This droplet formation occurs by creating localized hot spots within the bulk liquid, brought about by extreme temperatures and pressures. Consequently, ultrasonication is a robust and convenient method for producing micro and nanodroplets of LM. Typically, a bulk quantity of LM is placed in a vial containing a liquid medium, such as ethanol, and is subjected to ultrasonic treatment within a bath for several hours [32]. This process achieves a equilibrium between droplet breakup and coalescence, a balance that can be finely tuned by adjusting the fabrication process parameters. Sonication serves as a versatile tool, enabling the generation of LMDs and facilitating their surface modification with micro and nanoparticles.

3.5 Shearing

Mechanical agitation, mainly through shearing, is another efficient technique for producing small LMDs, with diameters typically ranging from tens of nanometers to a few micrometers. This process involves breaking down a bulk of LM by applying shear stress within a solvent, such as acetic acid. A shearing apparatus stretches the LM, forming elongated cylinders that ultimately fragment once they reach the Rayleigh–Plateau length limit. Tevis et al. introduced a straightforward method known as SLICE (shearing liquids into complex particles), which incorporates emulsion shearing while using the native oxide layer of EGaIn [39]. As depicted in Fig. 3E, the SLICE method offers flexibility in generating droplets of varying sizes by adjusting the shearing speed and the shear liquid employed. The resultant droplets can vary from a few nanometers to a few micrometers in diameter. SEM and TEM images of the shear-generated EGaIn micro and nanodroplets are shown in Fig. 3F.

3.6 Liquid metal marbles

Liquid marbles are commonly created by rolling a small aqueous droplet over a layer of powder, ensuring a consistent coating [31]. Consequently, various powder materials can be employed as liquid marble coatings, as they do not sink or amalgamate into the LM. Jeong et al. [46]. used Fe-coated LMs to magnetically control the flow and slug in microfluidic channels with varying geometries for electrical switching purposes. To achieve this, Fe-coated LMs were treated with HCl to acquire spherical shapes which were then collected via a syringe. The schematic of the procedure is shown in Fig. 3G. Depending on the type of coating used, the application can be modified. For instance, semiconductor-coated LM marbles can function as transistors or diodes [47]. In addition, using specific micro-/nanoparticles as coatings, such as ferromagnetic materials, facilitates the manipulation of these droplets or magnetic LM marbles using an external magnetic field [48].

4. Actuation principles and associated characteristics of liquid metals

GaLMs allows for significant and reversible alterations in a controlled manner when subjected to external stimuli. These external stimuli include variations in compressive or tensile mechanical stress, temperature, electric or magnetic fields, and light. That renders them excellent candidates for fabricating components that respond to stimuli. They are considered the most suitable metallic alloys for developing practical devices in stimulant-responsive systems. Among the five known metallic elements—Fr, Cs, Rb, Hg, and Ga—that are in the liquid state at or close to room temperature, only Ga is suitable for safe implementation in robotics for room temperature or near-human body temperature applications. Ga possesses the advantage of having virtually no vapor pressure at room temperature and exhibiting low toxicity, enabling safe handling [49]. Its melting point is slightly higher than the typical room temperature (~30 °C). However, the addition of other post-transition metals, such as In and Sn, allows for the

Table 1 Summary of the response characteristics of liquid metals [52–54].

Actuation principle	Mechanism	Characteristics	Challenges
Mechanical	• Fluidity	• High deformability • Fast recovery of conductive paths and highly stable healing ability	• Possibility of leakage LM out of the container • The mechanical force needs to be applied in a specific direction
Electric field	• Surface tension regulation	• Easy to control. • Large deformation and complaint movement style	• Consuming Ga or producing other unexpected chemicals
Magnetic field	• Lorentz force • Ferromagnetic force	• Non-contact manipulation, a significant driving force • On-demand movement • Change in reversible mechanical properties	• Require a rotating magnetic field or alternating magnetic field. • Most of the ferromagnetic particles have low solubility in GaLMs. • Relatively low efficiency
Chemical	• Marangoni effects • Surface tension regulation via triggering reaction taking place at the surface. • pH imbalance • Surface charge imbalance • Surface tension regulation via forming a galvanic cell	• Self-energy supply • Large-scale deformation • The surface of Ga-based LM can be functionalized with a variety of surfactants	• Less complexity • Require unique devices with aqueous chemical surroundings. • Less control of shape and movement

Temperature	• Thermal expansion • Phase transition	• Suitable for use in a wide variety of micro-environments • Construct different shapes easily in a large-magnitude (11 times) transformation. • Rapid response	• Difficulty in processing • More thermal expansion required
Optics	• Surface plasmonic effect • Light-induced heat generation • Light-induced chemical reactions (photo-induced electron-hole generation) • Bubble propulsion	• Soft platforms with surface for plasmonic applications • The stimuli can be applied remotely in a localized area. • Precisely switch the reaction on/off • Photothermal synergy • Non-contact	• Most are UV range plasmonic resonances for Ga-based LM droplets. • Only applying to the surface of LM, the core with a significant percentage of the total weight is not very useful. • Relatively low efficiency
Acoustic	• Acoustic radiation force	• Biocompatible	• Fabrication issues metal-based nano swimmers
Self-propulsion	• Chemical reaction	• Absence of external power source	• Uncertain interactions of motors

creation of Ga-based eutectic alloys, namely EGaIn (75% Ga, 25% In) and Galinstan (68.5% Ga, 21.5% In, and 10% Sn), respectively. Those eutectic alloys have a lower melting temperature than pure Ga, where their melting temperature is 15.7 °C or lower [50,51]. Moreover, GaLMs have exceptional adaptability as stimulus-responsive materials and synergize effectively with a wide range of organic and inorganic compounds, creating remarkable LM hybrids as intelligent materials.

The surface tension of GaLMs, crucial for accommodating various functional materials, can be controlled through different strategies [24]. Interestingly, the surface tension is significantly affected by an oxygenated environment, such as ambient air or water, where GaLMs undergo a responsive reaction, forming a self-passivating Ga-oxide skin [50]. This property offers remarkable possibilities for manipulating GaLMs through stimuli such as chemical reactions, pH variations, or electrochemical changes. Numerous fascinating on-demand responsive behaviors of Ga-based LMs and their hybrids have been documented, as presented in Table 1. The following sections will categorize and discuss the responsive behaviors of GaLMs, and their hybrids based on the type of stimulus applied. Moreover, recent advancements will be highlighted, which have enabled researchers to discover and elucidate their varied response behaviors to a range of external stimuli.

4.1 Mechanical actuation

LM refers to a type of metal that maintains its liquid state above its melting point. Its ability to flow is a direct result of its softness. At room temperature, the kinematic viscosities of water and GaLM are 1.01×10^{-6} and 2.5×10^{-7}—7.5×10^{-7} $m^2 \cdot s^{-1}$, respectively [55]. Consequently, its mobility can be compared to that of water. This characteristic significantly simplifies the operational aspects of working with LM, as it can be directly injected into specific containers. An even more remarkable is that its fluidity can conformally modify the LM surface shape along with the shape-changing of the peripheral substrate. Additionally, when an LM droplet becomes trapped in a closed loop filled with a solution, it can effectively pump the solution rather than just moving, making it suitable for transporting medicinal solutions. However, it is important to note that the presence of an oxide layer on the surface of LM can alter its surface tension and, subsequently, its flow characteristics. As a result, one can manipulate its deformation or flow properties by either forming or removing the oxide skin from the LM [56].

4.2 Electrical actuation

The second method is electrical actuation (voltage-driven). It does not require bulky pumps and may not even require direct contact with the LM, which is simple in implementation, control, and miniaturization. Additionally, these approaches can be employed for various scaling operations and provide precise control over the position and magnitude of the manipulation. In the following sections, each electrical actuation approach will be described in detail.

4.3 Electrocapillarity

In the 1870s, Lippmann [57] discovered the earliest known technique for modifying the interfacial tension of LM, which is the electrocapillarity technique. In the electrocapillarity method, an electrical potential is applied to an LM by a counter electrode in an inert electrolyte, causing a change in the effective interfacial tension of the metal due to the density of charge in the electrical double layer at the metal-solution interface. This electrical double layer acts like a capacitor (shown in Fig. 4A). To lower the capacitive energy at the surface, the surface area of the metal increases, resulting in a modification of the effective interfacial tension.

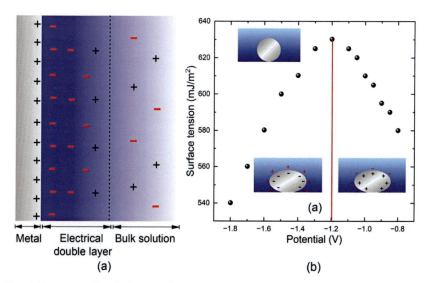

Fig. 4 Electrocapillary behavior of LMs in an electrolyte. (A) Electrical double layer. (B) The electrocapillary curve shows the surface tension of a liquid metal drop as a function of voltage.

Later, Lippmann [58] described the change for an ideally polarized electrode at constant composition as

$$-q = \frac{d\gamma}{dV},$$

(1)

where γ is the interfacial tension of the LM, V is the electrode potential, and q is the charge density at the interface. One can observe that an increase in charge density, regardless of whether it is positive or negative, leads to a reduction in the interfacial tension of the LM. This modification is described by an electrocapillary curve shown in Fig. 4B, which displays the change in interfacial tension as a function of potential [59]. As shown, at the peak of the electrocapillary curve, the "potential of zero charges" and the interfacial tension of the LM are at their maximum. The potential of zero charge value is a function of the metal and electrolyte.

The integration of the Lippmann equation (Eq. (1)) yields

$$\gamma(V) = \gamma_0 - \frac{1}{2}C(V - V_0)^2,$$

(2)

where γ_0 is the interfacial tension at the potential of zero charge, C is the capacitance at the double layer, and V_0 is the potential of zero charge.

Electrocapillary curves shown in Fig. 4A are valuable tools for characterizing the interfacial properties of mercury [60] and oxide-free liquid gallium [61] in a solution. However, it is essential to note that the curves can be influenced by the adsorption at the interface and the characteristics of the double layer. Therefore, they may vary depending on the specific electrolyte concentration and type being utilized [58].

Usually, there is confusion between the terms electrocapillarity and electrowetting in the literature. To clarify the difference between the terms (i.e., electrocapillarity and electrowetting), we follow the distinction made by Jackel et al. [62]. Electrocapillarity refers to the change in interfacial tension that occurs at the boundary between two fluids due to an applied electrical potential. Electrowetting refers to the change in wetting properties that occurs between a fluid and a separate material as a result of electrocapillarity. These two terms are often used interchangeably in the literature; however, they represent distinct phenomena. In the following section, the electrowetting phenomena will be addressed in detail.

4.3.1 Continuous electrowetting

A surface tension gradient can be generated across the LM surface when discrete drops of LM are placed in an aqueous solution, and an external

Micro- and nano-engineered liquid metals 143

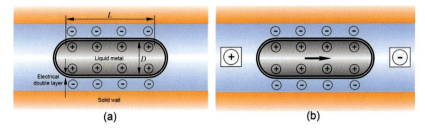

Fig. 5 Confinement of mercury in a microchannel, where electrodes are present on both ends of the microchannel; (A) in the absence of an applied voltage, the mercury plug remains stationary, and (B) application of a potential between the electrodes causes a gradient in interfacial tension resulting in the actuation of the metal plug.

electric field is applied. In accordance with the principles of electrocapillarity, this gradient arises due to a potential drop through the surrounding electrolyte (in the case of a plug of metal in a capillary, the thin layer of electrolyte between the metal and capillary walls), which can cause fluid motion inside of the channel without the need for direct contact between the LM and an electrode. Fig. 5 shows a mercury plug placed in a microfluidic channel filled with electrolytes. In the absence of an applied potential, the electrical double layer is evenly distributed on both sides of the drop. However, when a potential is applied across the ends of the microchannel, a potential drop occurs through the thin layer of liquid between the metal and the capillary walls, inducing an asymmetry in the electrical double layer across the drop. This asymmetry causes a difference in surface tension, resulting in the movement of the drop. This phenomenon is referred to as continuous electrowetting (CEW), and it is a direct consequence of electrocapillarity.

In the model of CEW for an incompressible Newtonian fluid confined to the capillary, the Navier-Stokes equations for interfacial forces are reduced to [63].

$$\frac{dp}{dx} = \frac{2}{d}\frac{d\gamma}{dx}, \quad (3)$$

where p is pressure, x is the dimension along the length of the capillary, and d is the inner diameter. Solving for the average velocity yields

$$v = -\frac{D}{6\mu}\left\langle \frac{d\gamma}{dx} \right\rangle, \quad (4)$$

where v is the average velocity of the fluid, μ is the viscosity of the LM, and $<d\gamma/dx>$ represents the average (over x) of the surface tension gradient.

The fluid physically flows towards the region of lower surface tension because it "attempts to wet more" the areas of lower surface tension. Furthermore, the voltage drop across the LM from the applied potential difference dictates the average interfacial tension difference. Substituting the potential gradient provides an estimate for the average velocity given as

$$v = -\frac{qD}{6\mu L}\Delta\varphi,$$

(5)

where L is the length of the drop, q is the charge in the electrical double layer at the interface between the electrolyte and LM, and $\Delta\varphi$ is the potential difference across it.

Eq. (5) implies that the velocity and the direction of the velocity are dictated by the externally applied potential, which can easily be controlled. Thus, in the absence of a potential, there is no gradient across the drop to drive the flow.

CEW enables the easy movement of plugs of mercury or gallium-based alloys within an electrolyte-filled channel; measures need to be taken to prevent adhesion of the oxide layer. Despite its potential to produce LM actuation with low power consumption, CEW has several limitations. For example, excessive voltages can cause hydrogen to form at the surface of the cathode via electrolysis, which can electrically isolate the electrode from the solution in microchannels, disrupting the operation of CEW. Although the use of AC voltages at higher frequencies can mitigate bubbles, some DC bias is generally necessary to achieve asymmetric motion of the fluid. Moreover, excessive potentials can cause the LM to split up inside microchannels. To overcome these challenges, electrowetting-on-dielectric and electrochemically-controlled-capillarity, which are other liquid actuation methods, can be used [64].

4.3.2 Electrowetting-on-dielectric

In this method, fluid modulation and actuation can be achieved beyond LM in electrolyte solutions, as charges at interfaces overcome the interfacial tension of any conductive solution, including organic materials containing dissolved electrolytes [65]. Fig. 6 illustrates the principle of the electrowetting-on-dielectric. As shown, a droplet of water on a conductive metal electrode can experience changes in its apparent contact angle at the surface

Micro- and nano-engineered liquid metals 145

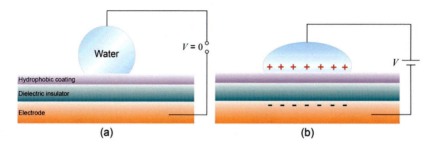

Fig. 6 Electrowetting-on-dielectric. (A) A conductive electrode coated with a thin dielectric and hydrophobic coating, with a water droplet placed on top. Upon release, the water droplet attains an equilibrium contact angle. (B) Upon application of a voltage between the electrode and the droplet, the charges migrate to the interface, thereby decreasing the effective contact angle.

of the electrode when a second electrode is inserted into the water, and a potential is applied between the electrode in the water and the metal substrate. Such a phenomenon is caused by charge rearrangement at the droplet's surface, which creates an imbalance at the three-point contact line between the vapor, liquid, and solid phases [64]. Mugele and Baret [57] provided a thorough review of electrowetting-on-dielectric phenomena.

In the absence of applied potential, the equilibrium contact angle is described by Young's equation [66], which is derived from a force balance at the three-point contact line and is given as

$$\gamma_{LV} \cos(\theta_Y) = \gamma_{SV} - \gamma_{SL}, \tag{6}$$

where γ_{LV} is the interfacial tension between the liquid and vapor phase, γ_{SV} is the interfacial tension between the solid and vapor phase, γ_{SL} is the interfacial tension between the solid and liquid phase, and θ_Y represents Young's contact angle. The application of a voltage to a liquid result in an increase in its surface area, leading to a decrease in its capacitive energy and a corresponding decrease in the apparent contact angle with both the vapor and solid phases, as described by

$$\cos(\theta) = \cos(\theta_Y) - \frac{\varepsilon\varepsilon_0}{2d_H\gamma_{LV}}(V - V_0)^2, \tag{7}$$

where θ represents the apparent contact angle at the microscopic scale, ε is the dielectric constant, ε_0 is the permittivity of the free surface, d_H is the thickness of the Helmholtz layer, V is the applied potential, and V_0 is the open circuit potential. A more extensive derivation can be found in the literature [57]. The angle of contact is determined by two factors: (1) the initial contact angle at equilibrium and (2) the ratio of capacitive

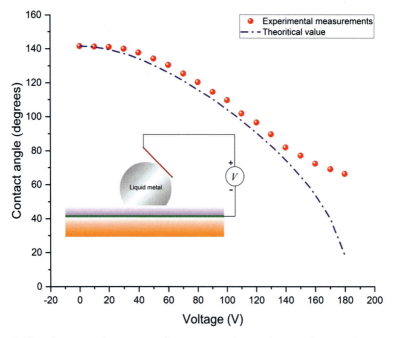

Fig. 7 The electrowetting curve of a mercury drop, where at large voltages, the contact angle begins to saturate and deviate from the theoretical curve. *Experimental and theoretical values are obtained from Ref. Z. Wan, H. Zeng, A. Feinerman, Reversible electrowetting of liquid-metal droplet, J. Fluids Eng. 129 (4) (2006) 388–394, https://doi.org/10.1115/1.2436582 [69].*

energy (which propels the droplet towards the substrate) to the surface tension of the liquid.

While EWOD has been traditionally used for manipulating aqueous droplets, its principles can also be applied to LMs [67,68]. Fig. 7 shows an electrowetting curve that indicates the change in contact angle with respect to voltage for mercury on a 600 nm thick Parylene (or perylene) film [70]. Similar outcomes have been observed for electrowetting using Galinstan [71] but under specific conditions to prevent the formation of the oxide layer, which introduces significant experimental challenges. The large interfacial tension of LMs necessitates a high voltage (>100s volts) to achieve a substantial change in contact angle, which is a notable drawback.

However, using EWOD to manipulate LMs has two primary drawbacks: (1) the ineffectiveness of EWOD on LMs that form surface oxides, and (2) the need for large voltages. Surface oxides present mechanical barriers to actuation, and while aqueous acids and bases can remove oxides, they are conductive and cannot be used with EWOD. The use of insulating fluids that can remove oxides from gallium alloys could help overcome this limitation. LMs also have high interfacial tension that must be overcome for deformation. While surfactants can decrease surface tension in water, achieving a comparable effect in LMs is challenging.

4.3.3 Electrochemically-controlled-capillarity

The capacity to selectively induce and eliminate the oxide surfactant permits straightforward injection and withdrawal within electrolyte-filled capillaries, a method known as "electrochemically controlled capillarity" (ECC). ECC results in a significant reduction in surface tension, propelling the LM into the channels. In theory, electrochemical reactions at surfaces have the potential to deposit species that aid in reducing surface tension. In typical electrocapillarity experiments, efforts are made to prevent electrochemical reactions, ensuring that changes in surface tension are solely attributed to electrical double-layer effects. However, when cathodic or anodic potentials become excessively high, bubbles or other reactive by-products may develop on the metal surface. This limitation constrains the voltage that can be applied in the realm of electrocapillarity.

While the electrochemical method shows promise, it is not without drawbacks. Injection of the LM leads to the formation of oxide, which requires using strong acids or bases to remove excessive oxide layers continually. Neutral electrolytes result in the build-up of oxide, which mechanically hinders the movement of the metal. It may be worth exploring other electrochemically active species that can lower surface tension without creating mechanical impediments to flow. Table 2 compares the advantages and disadvantages of using each of these techniques. In general, the methods that involve direct contact of the LM with an electrolyte require low voltages and low power to operate.

Table 2 Comparison of electrical-based actuation techniques of LMs [72].

Method	Voltage requirement	Electrolyte	Chemical reaction
Electrocapillarity	Low	Yes	No
Continuous electrowetting	Low	Yes	No
Electrowetting-on-dielectric	High	No	No
Electrochemically-controlled-capillarity	Low	Yes	Yes

4.4 Magnetic actuation

Remote application of magnetic fields to materials offers a means to avoid potential interference from invasive fuel-like environments, chemical reactions, and electrolyte splitting caused by electrodes. Magnetic fields, as a non-invasive and highly selective stimulus, exhibit minimal interactions with nonmagnetic materials and possess the ability to penetrate deeply into various materials, including biomaterials.

There are fundamentally two approaches to driving GaLMs using magnetic actuation. In the first approach, due to their high conductivity, LMs are capable of being moved by Lorentz forces induced by a changing magnetic field, generating a large eddy current in a NaOH solution [73]. The presence of an alkaline solution is essential to remove the oxide layer and prevent the LM droplet from sticking to the container. The Lorentz forces experienced by solid metal spheres lead to a horizontal force (which is in the same direction as the moving magnet) and a torque that makes it rotate in the opposite direction. While GaLMs are generally unresponsive to permanent magnetic fields at room temperature, they can still be effectively propelled by the Lorentz force due to their conductivity as fluids [74]. When an external alternating magnetic field is applied to an EGaIn LM, a range of physical phenomena can be observed, including exothermic behavior, controlled locomotion, electromagnetic levitation, and shape transformation of the LM [75], as shown in Fig. 8A. In Fig. 8B, it can be observed that a droplet of LM positioned above the center of a magnetic coil exhibits reversible "stand-up" deformation behavior when the alternating magnetic field is turned on or off. This behavior indicates that the eddy currents induced by the alternating magnetic field in the LM, in accordance with Lenz's law, generate a sufficient repulsive force for the LM droplet to overcome gravity.

The second approach involves the physical coating of ferromagnetic, semiconductors, and insulating materials, whether in the form of micro or nanoparticles, onto the surfaces of LMs to create what is popularly referred to in the literature as "liquid metal marbles" [24,25,30]. This term stems from "liquid marble," originally coined almost a decade ago to describe droplets of aqueous fluid enclosed within hydrophobic particles [31]. These liquid marbles exhibit characteristics similar to solid particles, but their structural form is primarily determined by surface tension, resulting in several distinctive properties. These properties include minimal contact areas with surfaces, which enable low friction rolling, superhydrophobic

Micro- and nano-engineered liquid metals

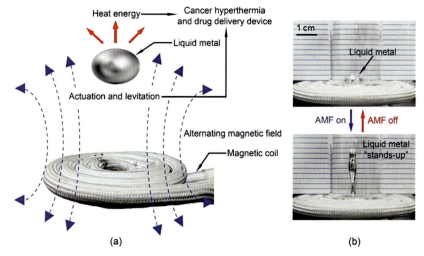

Fig. 8 (A) Alternating-magnetic-field (AMF) induced manipulation of the liquid metal blob. (B) Demonstration of AMF-modulated reversible deformation behavior of LM. *Adapted from Ref. Y. Yu, E. Miyako, Alternating-magnetic-field-mediated wireless manipulations of a liquid metal for therapeutic bioengineering, iScience 3 (2018)134–148. https://doi.org/10.1016/j.isci.2018.04.012, with permission obtained from Elsevier [75].*

interactions with other fluids, and the ability to merge or separate through self-healing encapsulation layers [76].

This technique has been extensively utilized to investigate the responsive capabilities of LMs to magnetic fields [77]. By introducing the ability to respond to magneto-stimuli, LM marbles can alter their motions and shapes when subjected to a magnetic field. This is achieved by utilizing magnetic materials as intermediaries to exert forces on the LM marbles under the influence of the magnetic field.

4.5 Optical actuation

High-frequency electromagnetic radiation, known as light, can be utilized to remotely stimulate materials. The use of light presents distinct benefits in adjusting external stimuli, such as the ability to regulate light intensity and frequency and precisely control the direction, position, area, and duration of irradiation. In the case of metals, the interaction between electromagnetic energy and the electronic band structure results in either re-emission or reflection of light, depending on the frequency of the incident light.

The Ga-based LM exhibits a plasma frequency, which represents the frequency at which the electron cloud oscillates. This frequency falls within the energy range comparable to ultraviolet (UV) light [78]. Consequently, the GaLM reflects light in the visible and infrared regions, behaving akin to a mirror. However, when photons with wavelengths shorter than UV light pass through the GaLM, localized surface plasmon resonances can occur due to the interaction between incident light and the free electrons [79]. Exploiting these soft LMs, there is an appealing opportunity to design plasmonic devices.

The generation of heat or thermal effects through light irradiation is another prevalent stimulus-responsive phenomenon that can be utilized to manipulate GaLM. Previous studies have shown that EGaIn-based nano-droplets undergo significant morphological transformations when subjected to light irradiation in aqueous dispersion [80]. Furthermore, when subjected to near-infrared (NIR) laser irradiation, the LM nano-capsules generate thermal energy and reactive oxygen species, leading to the destruction of the capsule structure.

Another intriguing application involves achieving photo-induced locomotion through the combination of photo-active semiconductors with LMs. LM marbles coated with WO_3 (Tungsten oxide) nanoparticles exhibit photocatalytic properties when exposed to light with a wavelength below 460 nm. Upon irradiation and subsequent photo-catalytic reaction in an H_2O_2 solution, the localized generation of oxygen bubbles induces a rolling force that propels the LM marble away from the evolving bubble region [81], as depicted in Fig. 9.

Furthermore, integrating GaLMs with other functional materials offers a promising avenue for enhancing the photo-induced effects. For instance, the incorporation of Mg into the EGaIn LM yields a notable 61.5% increase in photo-thermal conversion compared to the pure EGaIn LM, enabling efficient photo-thermal therapy for skin tumors [82]. It is believed that the formation of a new intermetallic phase, Mg_2Ga_5, within the LM marbles, contributes to this enhanced photo-thermal effect, although the precise mechanism remains incompletely understood.

4.6 Acoustic actuation

Acoustic cavitation is a phenomenon induced by ultrasound, which leads to the formation, growth, and eventual collapse of bubbles within a liquid. This process generates localized hot spots characterized by extremely high temperatures and pressures [16]. As a result, ultrasound serves as an efficient

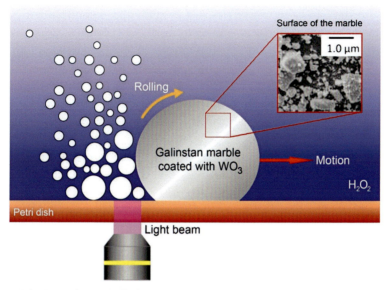

Fig. 9 Schematic diagram of a liquid metal marble coated with Tungsten oxide (WO$_3$) nanoparticles placed in an H$_2$O$_2$ solution irradiated by ultraviolet (UV) light. *The SEM image of liquid metal marble is reproduced from Ref. X. Tang et al. Photochemically induced motion of liquid metal marbles, Appl. Phys. Lett. 103 (17) (2013) 174104, https:// doi.org/10.1063/1.4826923, with permission obtained from AIP Publishing [81].*

and straightforward method for producing and propelling LM micro/nanoparticles.

Ultrasonication allows for adjusting the balance between nanoparticle breakup and coalescence by modifying temperature or introducing acid, which removes the native surface oxide layer. In addition, research has been conducted to fragment LMDs with and without oxide later in response to varying acoustic wave amplitudes, which may lead to practical LM inkjet developments [83]. Wang et al. [84]. employed LGNMs as nanomedicines for the photothermal therapy (PTT) of tumors, where they were propelled autonomously by an ultrasound field. The motion propelled by ultrasound provides an innovative approach for the contactless propulsion of LM nanodroplets.

4.7 Thermal actuation

The initial stimulus-responsive characteristic explored in a GaLM was volume expansion upon increasing temperature, as referenced by [85]. Galinstan was initially used as a substitute for toxic mercury in commercial thermometers and thermocouples. Ga exhibits the second widest liquid state range among all metals, spanning from 30 °C to 2400 °C. It possesses a distinctive thermal expansion coefficient that increases by several hundred percent after melting. The volumetric expansion coefficient of liquid Ga in the macroscopic state is 101.5×10^{-6} per °C within the temperature range of 30–977 °C [54]. This coefficient is higher than that of most solid metals but remains relatively lower compared to numerous other liquid substances and soft polymers.

The volumetric coefficient of thermal expansion (α) is given by

$$\alpha = \frac{1}{V}\left(\frac{\partial V}{\partial T}\right)_{p},$$

(8)

where the subscript p indicates that the pressure is held constant during the expansion, therefore, the expansion of Ga or GaLMs can be conspicuous when storing them in a small container, especially in nanosized containers [85].

Limited research has been conducted on the thermally driven responsive activities of GaLMs apart from their use in thermometers. This scarcity of studies can be attributed to Ga's low thermal expansion coefficient near room temperature, resulting in low sensitivity and minimal response of bulk LMs without confinement within small containers. However, researchers have successfully harnessed their thermally driven characteristics for practical applications by incorporating GaLMs into hybrid structures alongside other materials with stimulus-responsive properties.

The phase transition of LMDs, influenced by temperature, is associated with an abnormal volume expansion of the droplets [86]. This abnormal volume expansion causes the LMDs dispersed in silicone oil to encounter one another, resulting in the formation of a conductive path. The entire process is reversible, enabling the material to be used to create circuits with conductive insulation transformation functions. Sun et al. [87]. discovered that low temperatures prompt LMDs to change from an ellipsoidal to an amorphous shape. This is due to the strong impulse expanded force in liquid-solid phase transition in a dual fluid system composed of LMDs and aqueous solution, resulting in rapid, large-scale, and intense transformation. During the phase transition process, LMDs expand to form sword-like shapes, which demonstrates remarkable mechanical destruction and negligible biotoxicity, providing a new approach to tumor therapy.

4.8 Chemical actuation

The morphology of LM can be manipulated by adjusting the interfacial tension by modifying the chemical surroundings [88–93]. This manipulation holds potential for applications in flexible electronics, such as switches, resistors, capacitors, and reconfigurable antennas. Another effective approach to activate redox chemical reactions on LM surfaces is by constructing galvanic cells with conductive chemical surroundings. In this regard, regulating the surface tension of LMs can trigger their motion or deformation. Fig. 10 illustrates a typical galvanic cell design for the self-actuation of an LM, where the LM functions as an anode, and the copper electrode acts as a cathode [94]. The NaOH solution filling the channel between the reservoirs serves as an electrolytic bridge between the two half-cells. When an external electrical connection is established between the LM and the copper electrode, a spontaneous redox reaction occurs, with Ga undergoing oxidation at the anode and oxygen reduction taking place at the cathode. As a result, oxides form on the LM's surface, reducing the localized interfacial tension and inducing LM flow into the channel.

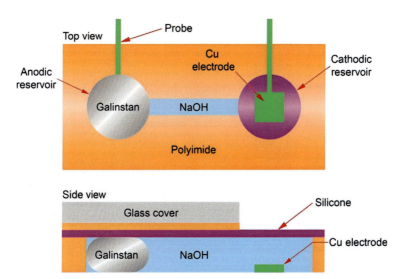

Fig. 10 Schematic illustration of the mechanism for the actuation of liquid metal via redox reaction by forming a galvanic cell system. Galinstan was wetted to a copper probe in a cylindrical "anodic" reservoir. An identical "cathodic" reservoir is on the opposite side of a fluidic channel and contains an embedded copper electrode. The reservoirs and the intermediate channel are filled with 0.25 M NaOH solution as an electrolyte [94].

On breaking the external electrical connection, the redox reaction ceases, and the Ga–oxides dissolve in NaOH, causing the LM to flow back from the channel. By creating a galvanic cell, significant deformation or displacement can be achieved.

Similarly, shapeshifting and fractal phenomena can be achieved in acid environments by introducing ions like Cu^{2+} to form a Cu–Ga galvanic cell, triggering displacement reactions on the LM's surface and inducing imbalanced interfacial tension [90]. To better understand and achieve accurate on-demand mobility control of LMDs, it is essential to explore and quantify the influences of different electrolytes, including acidic, alkaline, and neutral solutions, on the oxidation formation on LM surfaces [88].

4.9 Self-propelling actuation

Extensive research has been conducted on the development of LM systems capable of self-actuation, where the energy derived from spontaneous chemical reactions is converted into mechanical actuation, enabling the autonomous movement of LMDs [95,96]. For instance, it has been demonstrated that LM objects can efficiently self-actuate by placing a small Al flake on their surface as fuel. The self-propulsion mechanism relies on the force generated by bubbles resulting from the Al reaction in the solution and the surface tension imbalance caused by the bipolar electrochemical reaction facilitated by the alloyed metal. Notably, Al has been extensively studied as a fuel for actuation due to its high alloying capability with LM [53,95,97]. This self-actuation ability of LMs offers the potential for new techniques beyond soft robot applications, such as self-powered machines, tuneable RF, and microfluidic devices.

The Marangoni effect, empowered by the electrochemical reaction between Al and EGaIn and the hydrogen evolution at the reaction sites, is the primary reason for the results observed in this study, as shown in Fig. 11A. To initiate this effect, a droplet of EGaIn is dropped into a NaOH solution, and a small Al sheet is placed next to it, causing the LM to break into the passivation layer on the Al sheet and create an alloy of Al and LM. When this alloy (Al dissolved in LM) is collected and injected back into the NaOH solution, a stream of tiny LMDs is generated, which move in different directions like running motors. Similarly, Yuan et al. [53]. developed a method to create self-powered LM motors with Brownian motion in an alkali electrolyte by dissolving 1% of Al in $GaIn_{10}$ (Ga 90%, In 10%), as depicted in Fig. 11B. The hydrogen bubbles also enable the main driving force in this case.

Micro- and nano-engineered liquid metals

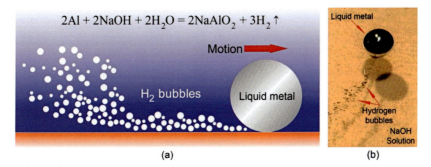

Fig. 11 (A) Self-propulsion of liquid metal, where the H$_2$ bubbles push the liquid metal forward. The motion is a combination of rolling and translating mechanisms. (B) Image showing H$_2$ bubbles arising from the bottom of the droplet. *Adapted from Ref. B. Yuan, S. Tan, Y. Zhou, J. Liu, Self-powered macroscopic Brownian motion of spontaneously running liquid metal motors, Sci. Bull. 60 (13) (2015) 1203–1210, https://doi.org/10.1007/s11434-015-0836-6, with permission obtained from Elsevier [53].*

4.10 Multi-stimuli actuation

Investigating the synergistic effects of multi-factor stimulation presents an intriguing area of exploration in designing intelligent materials for LM-based units. The advancement of LM systems capable of responding to multiple stimuli holds significant appeal due to the potential for heightened responsivity and improved precision.

In the context of achieving multi-stimuli responsive behavior in LMs, a vital consideration is the ability to manipulate them through the regulation of surface properties and interfacial tension. The interplay between surface tension modulation, density, and gravity significantly impacts stimulus-response behavior. While gravity remains relatively constant during the stimulus–response process, the density of LMs and their hybrids plays a crucial role in the responses. The initial design of density can be achieved by selecting suitable metals for mixing. Moreover, the incorporation of various nano- and micro-particles can alter the density based on the wetting and alloying properties of the added components [25].

In addition to the considerations regarding gravity and density, limited research has been conducted thus far on the investigation of multi-stimuli responsive behavior in LMs and their hybrids. One notable example is the achievement of precise control and locomotion of LM hybrids through the combined influence of magnetic and electrical stimuli [98]. In this study, LM hybrids were created by incorporating copper-iron magnetic nanoparticles into the LM composite. Consequently, these hybrids exhibited electrochemically driven motion and rapid responses to an applied magnetic field. An advanced behavior showcased the interplay between electric and magnetic fields, enabling precise manipulation of actuation direction within a cross-linked channel and climbing-type locomotion. This design holds promising potential for the development of untethered and remotely controlled soft tools for surgical procedures and drug delivery. Overall, the exploration of applying multiple stimuli to LMs opens new avenues for expanding their regulation range and responsivity while also offering unexpected responsive properties.

5. Applications of liquid metals in thermal engineering
5.1 Phototherapeutic

Gallium (Ga)-based LMs (GaLMs) exhibit highly efficient light-to-thermal energy conversion, leading to recent advancements in photothermal and photodynamic applications for anti-cancer therapies. Consequently, an array of phototherapies, including PTT and photodynamic therapy (PDT), has seen rapid development due to their exceptional efficacy in localized tumor treatments [99,100]. PTT, for instance, functions by destroying tumors through the conversion of light into thermal energy, leading to a rapid increase in temperature [72,101]. This, in turn, necessitates the use of photothermal agents capable of converting light energy into hyperthermia for the eradication of tumor tissues [102,103]. On the other hand, PDT involves photosensitizers that are activated by light at specific wavelengths. Consequently, recent approaches in phototherapy have proposed the use of optimized photosensitive materials, particularly for efficient light-to-heat or energy conversion through photochemical reactions.

In this context, LM emerges as a versatile photothermal and photosensitizing agent, showcasing distinctive attributes like metal-like conductivity, ease of fabrication [16], and remarkable shape adaptability [104]. Notably, there has been a significant upsurge in research exploring the potential of Gallium-based LM in anticancer therapeutics over the past decade (Fig. 1).

The process of converting absorbed light into heat can be classified into three distinct phases (depicted in Fig. 12): when a metal nanostructure is exposed to light, (1) electronic excitation initiates swift surface heating through electron-electron coupling, (2) subsequent cooling takes place through the transfer of energy between electrons and lattice phonons, and

Micro- and nano-engineered liquid metals 157

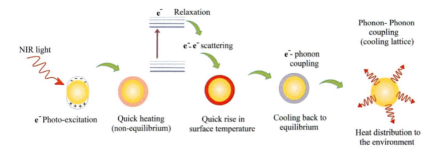

Fig. 12 Scheme of photo thermal effect mechanism under near-infrared laser irradiation. *Reproduced from Ref. J. Kadkhoda, A. Tarighatnia, J. Barar, A. Aghanejad, S. Davaran, Recent advances and trends in nanoparticles based photothermal and photodynamic therapy, Photodiagnosis Photodyn. Ther. 37 (2022) 102697, https://doi.org/10.1016/j.pdpdt.2021.102697, with permission obtained from Elsevier [105].*

(3) this is followed by phonon–phonon coupling, which leads to the dissipation of heat into the surrounding environment. As a result, the irradiation of metal nanoparticles induces a significant temperature difference between the particle's surface and its surroundings, leading to a profound increase in local temperature.

Based on this mechanism, PTT utilizes excess heat energy generated from absorbed light sources by photothermal agents. Photothermal conversion efficiency (PCE) is a primary evaluation index for photothermal efficacy and is calculated by

$$\eta = \frac{hS(T_{max} - T_{surr}) - Q_{dis}}{I(1 - 10^{-A_\lambda})}, \tag{9}$$

where η is the PCE, T_{max} is the maximal temperature of the system, and T_{surr} is the surrounding temperature, so $(T_{max} - T_{surr})$ is the temperature change. I is the laser power, A_λ is the absorbance at a specific wavelength of the laser, h is the heat transfer coefficient, and S is the surface area of the container. Then, hS can be calculated as follows

$$hS = \frac{mC_p}{\tau_S}, \tag{10}$$

$$\theta = \frac{\Delta T}{\Delta T_{max}} = \frac{T - T_{surr}}{T_{max} - T_{surr}}, \tag{11}$$

where m is the mass of the system, C_p is the heat capacity of solvent, τ_S, and the time constant is measured from the cooling stage of the photothermal curve. The dimensionless parameter θ is calculated by the ratio of ΔT to

ΔT_{max}, where T is the sample temperature [106]. A material with high PCE can be considered as a suitable candidate photothermal agent. Table 3 summarizes modulated PCEs of various Ga-based LM materials by changing the wavelength and power.

5.2 Thermal switches

For many years, the thermal management community has sought after devices that can actively regulate heat flow. The demand for thermal control has grown more pressing, especially with the increased power density in devices, leading to localized heat fluxes as high as $1 \, kW \cdot cm^{-2}$. Thermal switches, which can alternate between high and low thermal conductance, provide the means to partition and actively manage heat flow pathways.

Recent advancements in electronic devices have led to increased power density (power-to-volume ratio) and specific power (power-to-weight ratio) in both stationary and mobile systems. The shift towards replacing traditional mechanical systems with smaller electrical counterparts has created a demand for lighter and more compact electronic devices [115]. Nevertheless, the ability to manage heat flow from hotspots is a critical factor in the design and operation of electrical systems, especially on a millimeter scale. In the case of power electronics, heat generation can exhibit temporal and spatial irregularities, leading to temperature spikes and gradients that can adversely affect system performance and reliability.

While passive systems using thermal buffers based on phase-change materials can partially mitigate these issues [116], technologies that enable active thermal switching offer increased flexibility in designing and operating electro-thermal systems [117]. A thermal switch, analogously defined to electronic components [118], acts as a thermal circuit element that controls the effective thermal conductivity between hot and cold reservoirs, enabling the regulation of heat transfer—switching it "ON" or "OFF." Thermal switches facilitate active thermal management for high-power electronics, addressing the challenge of temperature equalization, particularly when sharing power among parallel-connected devices. They can also assist in lowering the temperature of temperature-sensitive devices located in close proximity to heat sources [119]. Furthermore, a versatile and durable thermal switch would open doors to additional thermal circuit elements, such as thermal diodes and thermal rectifiers [117].

Yang et al. [120]. developed an LM-based thermal switch technology through integration and heat transfer characterization with gallium nitride

Table 3 Photothermal conversion efficiency (PCE) [106].

Gallium-based material	Irradiated laser wavelength (nm)	Laser power	Photothermal conversion efficiency (%)	References
EGaIn nano-rice	808	2.0 $W \cdot cm^{-2}$	36.7	[107]
EGaIn nanorod	808	2.0 $W \cdot cm^{-2}$	28.8	[107]
EGaIn nanosphere	808	2.0 $W \cdot cm^{-2}$	33.3	[107]
EGaIn	808	1.5 W	26.19	[108]
Mg-GaIn	808	1.5 W	42.18	[108]
Bare GaNPs	808	1.08 W	32.6	[82]
Ga@reduced graphene oxide core-shell NPs	808	1.08 W	42.4	[82]
EGaIn-based nano-platform by silica nanoshell	1,064,808	1.5 $W \cdot cm^{-2}$ 2.0 $W \cdot cm^{-2}$	22.43 17.14	[109]
Bare EGaIn	1,064,808	1.5 $W \cdot cm^{-2}$ 1.5 $W \cdot cm^{-2}$	14.12 10.21	[109]
Ga nanorod	808	1.5 W	32.72	[110]
Ga nanosphere	808	1.5 W	25.33	[110]
EGaIn nanorod	808	1.5 W	25.27	[110]

(continued)

Table 3 Photothermal conversion efficiency (PCE) [106]. (*cont'd*)

Gallium-based material	Irradiated laser wavelength (nm)	Laser power	Photothermal conversion efficiency (%)	References
EGaIn nanocapsule	785	1 W	52	[111]
EGaIn	808	1.0 W·cm^{-2}	28.3	[112]
EGaIn with mesoporous silica shell	808	1.0 W·cm^{-2}	51.5	[112]
Core-shell GaIn@Pt NPs	1064	1.0 W·cm^{-2}	~39.08	[113]
Bare GaIn NPs	1064	1.0 W·cm^{-2}	15.50	[113]
EGaIn-Au NPs	808	1.5 W·cm^{-2}	22.58	[114]
EGaIn NPs	808	1.5 W·cm^{-2}	7	[114]

(GaN) electronic devices. Fig. 13A shows a schematic of the thermal switch concept [19], while Fig. 13B shows a thermal switch integrated with a single GaN device mounted on a PCB. When the thermal switch is in the ON mode, the LM droplet ($k_{Galinstan} \approx 16.5 \text{ W·m}^{-1}\text{·K}^{-1}$) enables efficient heat transfer from the GaN device (heat source) to the coolant (heat sink) at the front. In the OFF mode, the LM droplet is moved away, breaking the efficient heat transfer to the coolant. The channel is filled with a low thermal conductivity electrolyte solution ($k_{electrolyte} \approx 0.6 \text{ W·m}^{-1}\text{·K}^{-1}$) or electrolyte vapor ($k_{vapour} \approx 0.03 \text{ W·m}^{-1}\text{·K}^{-1}$). They showed that the thermal switch could enable GaN device for a variety of power dissipations, resulting in lower thermomechanical stresses and higher long-term reliability.

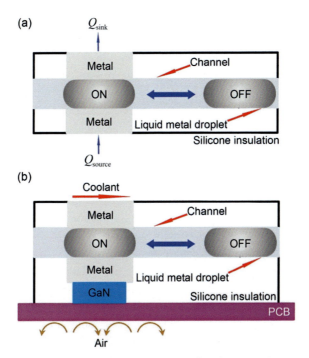

Fig. 13 Schematic of the thermal switch integrated with a GaN electronic device. (A) Liquid metal droplet can move from left (ON) to right (OFF) in the channel. (B) The heat from the GaN device (Q) is removed from the top via forced convection to cooling water and from the bottom via natural convection to ambient air. *Redraw based on Ref. T. Yang et al. An integrated liquid metal thermal switch for active thermal management of electronics, IEEE Trans. Components, Packag. Manuf. Technol. 9 (12) (2019) 2341–2351, https://doi.org/10.1109/TCPMT.2019.2930089 [120].*

5.3 Photothermal actuators

A photothermal actuator is a device that converts light energy into mechanical energy by harnessing thermal energy. In comparison to other types of photo-actuators, photothermal actuators offers a straightforward structure and easier implementation [121]. Moreover, through careful structural design, photothermal actuators can achieve localized actuation without impacting the overall state of the device [122]. Additionally, photo-actuators can deliver precise and controllable actuation by adjusting parameters such as the wavelength and intensity of the light source [123]. As a result, photothermal actuators have garnered increasing attention and are widely applied in fields like flexible robotics [124] and biomimetic systems.

Photothermal actuators typically employ photothermal fillers such as carbon nanotubes (CNTs), graphene, or metal nanocrystals to convert light energy into thermal energy. For instance, Xu et al. developed a temperature-controlled modulus CNT nanocomposite yarn photo-actuator [123]. This actuator relies on CNTs as the photothermal material and is driven by an infrared laser. However, due to the remarkable thermal conductivity of CNTs, it can't perform localized photo-actuation.

In contrast, metallic photothermal fillers like gold nanocrystals are well-dispersed and isolated within the matrix. Even in small quantities, they avoid long-range heat transfer and facilitate localized photo-actuation [122]. For example, Sun et al. [125]. introduced gold nanocrystals into liquid crystal elastomers. The heat generated by the gold nanocrystals under the laser beam is selectively transferred to submicron-sized particles. This, in turn, enables photothermal control within the liquid crystal elastomer, leading to reversible and irreversible shape changes.

Nevertheless, the majority of photothermal fillers consist of inorganic rigid particles, which can significantly impact the flexibility of the actuator's substrate and lead to a substantial reduction in the photo-actuator's response speed. In contrast, LM, an emerging flexible material, amalgamates the fluidity of a liquid [126] with the electrical conductivity [127] and thermal conductivity [128] of a metal. Furthermore, similar to gold nanocrystals [129] and silver nanocrystals [130], LM particles exhibit photothermal effects when exposed to NIR laser [131]. Moreover, owing to the flexibility of LM, it can seamlessly integrate with the polymer matrix as a filler, accommodating various deformations without compromising the photo-actuator's response speed. Consequently, LM holds excellent potential for the development of photothermal actuators. However, it is worth noting that while the photothermal properties of LM particles are predominantly harnessed in medical applications [107], there have been relatively few studies focusing on the utilization of LM microspheres in photothermal actuators.

5.4 Energy harvesting

Efficiently harnessing solar energy is a priority to mitigate the energy crisis. In this regard, renewable energy plays a vital role in the rising energy demand for energy and reduces greenhouse gas emissions. Solar radiation is a source of renewable energy from the sun, and it can generate electrical and thermal energy using appropriate harvesting techniques. Presently, solar energy harvesting primarily relies on two approaches: solar photo-voltaic technology, which directly converts solar photon energy into electricity through the photovoltaic effect at semiconductor interfaces, and photothermal technology. In photothermal technology, solar radiation is transformed into heat energy using solar-to-heat materials, allowing for the utilization of a broader spectrum of sunlight, which results in superior conversion efficiencies. Photothermal conversion technology has one of the smallest carbon footprints among all photon energy harvesting methods [132], aligning with the dual-carbon concept of carbon peaking and carbon neutrality [133]. Regarding sustainable photothermal systems, the versatility and customizability of photothermal materials are advantageous for advancing and popularizing photothermal technology [134].

From the perspective of ideal solar-thermal structural engineering, the development of photothermal enhancers with selective and strong broad-band light absorption across the entire solar spectrum while minimizing heat energy radiation loss becomes pivotal for the efficient utilization of solar-to-thermal energy. Ever since the introduction of eutectic gallium/indium (EGaIn), an LM that remains in a near-room-temperature state, this alloy material has shown immense promise for applications in high-efficiency thermal management and solar-thermal-driven power generation [134]. Additionally, LMs offer an ideal starting point for the development of highly efficient photothermal materials, thanks to their intriguing properties. These properties include the ability to accommodate higher heat flux density, expedite heat energy transfer, and serve as a reliable platform for thermal energy storage within photothermal materials. Furthermore, when combined with interface engineering, creating stable polymer/LM dispersoids becomes particularly crucial. This is essential for achieving durable light-to-heat conversion performance, ensuring excellent flexibility, and maintaining long-term structural reliability in photothermal polymer/LM composite materials.

Yang et al. [134]. developed an effective dual-interfacial integration strategy to enhance the photothermal properties of EGaIn nanodroplets. They achieved this by systematically encapsulating them with a polydopamine coating of adjustable thickness and a reduced graphene oxide shell. The schematic guideline for converting the photothermal absorber is shown in Fig. 14. This approach has improved the heat utilization efficiency and solar light absorption of these nanodroplets. Benefitting from the construction of a stepped micro-grating architecture and the even dispersion of inclusions, the resulting PVA-based absorber exhibits quick and uniform heat transfer characteristics, minimal heat energy loss, robust broadband light absorption, and high structural reliability. Incorporating dual light-absorption shells and retroreflection for light trapping collectively led to outstanding photothermal conversion of 89.4% and superior broadband solar absorption > 94.9%. Significantly, this study showcases a promising application of

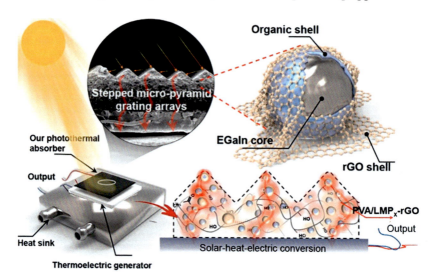

Fig. 14 Synthetic roadmap to convert bulk LM to LMP-rGO nanodroplets and the fabrication of PVA-based photothermal absorber featuring the 3-D stepped micropyramid grating array surface. *Adapted from Ref. S. Yang, Y. Zhang, J. Bai, Y. He, X. Zhao, J. Zhang, Integrating dual-interfacial liquid metal based nanodroplet architectures and micro-nanostructured engineering for high efficiency solar energy harvesting, ACS Nano 16 (9) (2022)15086–15099. https://doi.org/10.1021/acsnano.2c06245, with permission obtained from ACS Publications [134].*

the 3D micro-grating absorber in solar-driven heating, wide-angle solar energy harvesting, and clean power generation under solar irradiation conditions. These developments are expected to contribute to the high efficiency harvesting and utilization of solar energy.

5.5 Microfluidic thermal management

Thermal management solutions known for their exceptional cooling performance, including techniques like spray cooling, jet impingement, two-phase cooling, cold plates, and PCM, have proven effective in efficiently dissipating heat from hot spots. However, these conventional approaches fail to encounter challenges when dealing with temperature variations among multiple electronic devices within the same module. Moreover, these techniques are inherently complex to be integrated into circuits and chips, significantly cooling the hot spots in the extremely confined space inside the 3D stacked packages. Therefore, effective integrated/embedded thermal management methods have gained increasing interest to overcome the physical constraints in electronic devices with shrinking sizes, which cannot be tackled by conventional cooling technologies. Thus, microfluidic cooling systems based on semiconductor fabrication technologies enable miniaturization of the system size and easy integration into electronic devices. These offer promising solutions to on-chip or in-chip cooling for 3D IC packaging [135].

Microfluidic thermal management methods typically fall into two categories: continuous liquid-phase and discrete liquid-phase systems. Discrete droplet cooling methods, driven by surface-tension modulation, offer an energy-efficient option for integrated on-chip cooling systems. These systems achieve high coolant flow rates by periodically adjusting the surface tensions of the front and back menisci of each discrete microdroplet in response to externally applied fields [125]. Various methods of surface-tension modulation for droplet actuation have been proposed (as discussed in Section 5).

In 2005, Mohseni [136] introduced the concept of active thermal management based on the CEW principle. This innovative approach involved the discrete actuation of LM droplets, guiding them along electrode patterns through controlled modulation of surface tension by switching the electrodes on and off in a specific sequence. This method effectively demonstrated the potential of CEW as an efficient cooling technique for high-power-density electronic packages and the mitigation of hot spots in electronic devices. In a related technique within the realm of digital microfluidics, Mohseni and Baird [137] coined the term "digitized heat transfer" (DHT) to describe this cooling method. In DHT, thermal energy, or waste heat, is transported out of the device in a discrete manner, with individual droplets being actuated sequentially by electrodes. This approach fundamentally differs from conventional cooling techniques that rely on heat conduction through solid conductive materials and convection via continuous liquid flow.

Because of its fluidic characteristics at room temperature and high thermal conductivity, mercury has frequently been employed to illustrate the electrical control of discrete droplet motion using CEW and DHT for compact thermal management applications [135]. Nonetheless, mercury is a toxic heavy metal, making it unsuitable for general cooling applications. Liquid alloys, characterized by their low melting points near or below room temperature and favorable thermal properties, present a viable alternative to replace mercury in CEW-based thermal management systems. Therefore, CEW-based cooling techniques can be further advanced without the associated environmental and health risks. Zhu et al. [138]. utilized a Galinstan droplet positioned on a hot spot as part of a CEW-based cooling system, as shown in Fig. 15A. In their designed system, the Galinstan droplet, actuated by a square wave signal, functioned as a "soft pump" (Fig. 15B). The square wave signal created a surface tension gradient across the droplet, inducing a Marangoni flow in its vicinity. Consequently, the flow rate of the coolant around the Galinstan droplet increased due to the Marangoni flow. Simultaneously, the excellent thermal conductivity of the Galinstan droplet facilitated the transfer of waste heat from the hot spot to the coolant, as shown in Fig. 15C. This, in turn, enhanced the overall single-phase heat transfer performance.

5.6 Thermal interface materials

In recent years, there has been a rapid development of flexible electronic devices, such as wearable devices, flexible pressure sensors, flexible batteries, and flexible supercapacitors. Consequently, the thermal management materials used in these flexible devices must also possess similar flexibility to accommodate equipment deformation and maintain effective heat dissipation performance. Polymers, known for their flexibility, are commonly used as the base material for Thermal Interface Materials (TIMs). However, their thermal conductivity is typically relatively low, around $0.1 \text{ W·m}^{-1}\text{·K}^{-1}$, which may not meet the heat dissipation requirements driven by the rapid increase in electronic equipment power. Incorporating LM into polymers significantly

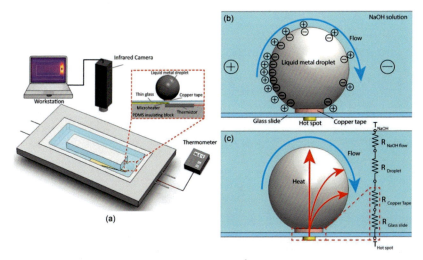

Fig. 15 (A) Assembled schematics of liquid metal-based cooling system. (B) Schematic of liquid metal droplet surface charge distribution when an electric potential is applied between the electrodes. The generation of harmonic Marangoni flow is enabled by a continuous electrowetting effect at the surface of the liquid metal droplet upon applying a square wave DC signal (C) Schematic and equivalent thermal circuit of heat dissipation through the liquid metal droplet, where T is temperature, and R is thermal resistance. *Reproduced from Ref. J.Y. Zhu, S.-Y. Tang, K. Khoshmanesh, K. Ghorbani, An integrated liquid cooling system based on Galinstan liquid metal droplets, ACS Appl. Mater. Interfaces 8 (3) (2016) 2173–2180, https://doi.org/10.1021/acsami.5b10769, with permission obtained from ACS Publications [138].*

enhances thermal conductivity while preserving their inherent flexibility [139]. LM-based polymer composites exhibit high stretchability, toughness, electrical insulation, and thermal conductivity, making them promising materials for applications in wearable electronics, soft robotics, thermal management, and more.

Among the key factors influencing the heat dissipation of microchips, the thermal contact resistance (TCR) between the microchips and the heat sink becomes prominent [140]. It is a well-established fact that when two flat solid surfaces come into mechanical contact, the presence of surface irregularities leads to the formation of air gaps, resulting in an increase in TCR at the mating surfaces. As a result, applying TIMs becomes necessary to fill these gaps, thereby reducing TCR and elevating the heat dissipation capabilities of electronic devices. To eliminate such air voids and reduce TCR, TIMs have been introduced between these interfaces. Conventional TIMs based on silicone have found widespread use due to their ability to readily adhere to rough interfaces [141]. However, the thermal conductivity of these

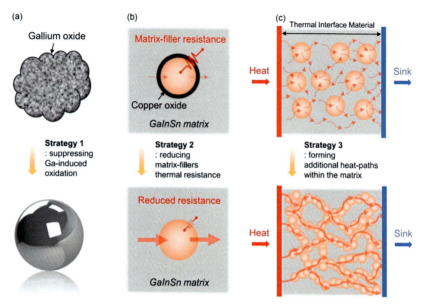

Fig. 16 Concept of the strategies for enhancing the thermal performance of LM-based TIMs. Schematics of three strategies: (A) suppressing Ga-induced oxidation of the LM matrix, (B) reducing interfacial thermal resistance between the matrix and fillers, and (C) forming additional heat-path networks within the matrix *Reproduced from Ref. S. Ki et al. Gallium-based liquid metal alloy incorporating oxide-free copper nanoparticle clusters for high-performance thermal interface materials, Int. J. Heat Mass Transf. 170 (2021) 121012, https://doi.org/10.1016/j.ijheatmasstransfer.2021.121012, with permission obtained from Elsevier [142].*

TIMs typically remains low ($< 9 \text{ W} \cdot \text{m}^{-1} \cdot \text{K}^{-1}$), owing to the intrinsic low thermal conductivities of the base materials ($\sim 0.5 \text{ W} \cdot \text{m}^{-1} \cdot \text{K}^{-1}$), even when incorporating conductive additives, such as silver (Ag), CNTs, aluminum oxide (Al_2O_3), and zinc oxide (ZnO) particles [142]. Ki et al. [142] presented three strategies to improve the thermal performance of LM while maintaining its fluidity, shown in Fig. 16. These strategies include (i) suppressing Ga-induced oxidation during the particle internalization processes in the GaInSn matrix, (ii) reducing interfacial thermal resistance between the matrix and fillers, and (iii) creating additional heat pathways within the GaInSn matrix through nanoparticle clusters.

5.7 Composite materials

Through the integration of LM microdroplets into a soft elastomer, Bartlett et al. [128]. successfully engineered a LM embedded elastomer (LMEE) that has exceptional stretchability (Fig. 17A). An additional method to further

Micro- and nano-engineered liquid metals 169

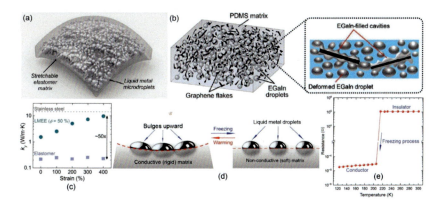

Fig. 17 (A) Schematic showing the liquid metal microdroplets dispersed in a flexible and stretchable elastomer matrix. (B) Schematic of graphene flakes as the second filler. (C) Thermal conductivity values as a function of strain for the pristine elastomer ($\Phi = 0\%$) and the LMEE ($\Phi = 50\%$). (D) The phase change of LM droplets in response to freezing and warming. (E) Experimental measurements for resistance as a function of temperature were obtained from Wang et al. [143]. *(A)–(C) Reproduced from Ref. K. Akyildiz, J.-H. Kim, J.-H. So, H.-J. Koo, Recent progress on micro- and nanoparticles of gallium-based liquid metals: from preparation to applications, J. Ind. Eng. Chem. 116 (2022) 120–141, https://doi.org/10.1016/j.jiec.2022.09.046, with permission obtained from Elsevier [37].*

increase the thermal conductivity of composite materials could be introducing secondary highly thermally conductive filler materials (such as graphene) to create thermal pathways to enhance the thermal conductivity, as shown in Fig. 17B. As depicted in Fig. 17C, an LM volume fraction (Φ) of 50% resulted in a remarkable enhancement of thermal conductivity under stress-free conditions, increasing by approximately 25-fold (4.7 ± 0.2 W·m^{-1}·K^{-1}) in comparison to the base polymer (0.20 ± 0.01 W·m^{-1}·K^{-1}). Even more strikingly, thermal conductivity surged nearly 50 times (9.8 ± 0.8 W·m^{-1}·K^{-1}) after applying strain. This exceptional enhancement was made possible by the unique thermal-mechanical interplay that integrated elongated LM inclusions within the soft elastomer, providing thermally conductive pathways and amalgamating thermal and mechanical attributes. Moreover, the researchers demonstrated an elastomer composite consisting of LM in which the microstructure of liquid inclusions could be tailored for various LM loadings. These LM-programmed elastomers were developed from a composite originally featuring spherical LMDs dispersed throughout a thermoplastic elastomer matrix. By manipulating the inclusions into controlled microstructures in stress-free configurations and thermally relaxing the thermoplastic polymer matrix, thermal conductivity reached an impressive $13.0\ W\cdot m^{-1}\cdot K^{-1}$, marking a seventy-fold increase compared to the polymer matrix [144].

Additionally, the thermo-responsive phase transition characteristics of LMs can also be leveraged to develop a polymer composite capable of undergoing a reversible insulator-to-metal transition in response to temperature changes. Remarkable reversible changes in resistivity (over 4×10^9 times) can be achieved by controlling the temperature during the warming and freezing cycles of this composite. This reversible insulator-to-metal transition in response to temperature changes primarily arises from the negative thermal expansion of GaLMs during the liquid-to-solid transition. Moreover, LMDs can protrude from a rigid silicone shell upon freezing, establishing conductive pathways by connecting with each other, as depicted in Fig. 17D. When the temperature increases, the droplets revert to a liquid state and retract within the soft silicone shell, breaking the conductive paths. This temperature-stimulated exploration of an insulator-to-metal (Fig. 17E) transition opens significant possibilities for various applications, including stretchable switches, semiconductors, temperature sensors, and resistive random-access memories. Consequently, we propose that GaLM hybrids can play a significant role in advancing temperature modulation and heat-driven smart behavior in soft robots.

5.8 Microencapsulated phase change materials

PCMs are highly efficient heat storage materials capable of absorbing and releasing substantial thermal energy with minimal temperature fluctuations during the phase change process. This exceptional characteristic makes them widely applicable in energy storage and temperature regulation domains [145]. Based on their chemical composition, PCMs are typically categorized into organic PCMs and inorganic PCMs [146]. Organic PCMs encompass a broad spectrum, including paraffin, fatty acids, their eutectic blends, esters, sugars, sugar alcohols, and other organic compounds. However, their thermal conductivity is generally low and prone to leakage and volatilization during use [147]. Inorganic PCMs currently consist of hydrated salts, molten salts, metals, and alloys. In comparison to organic PCMs, inorganic PCMs offer a higher volumetric heat storage density owing to their greater latent heat and density [148]. Nevertheless, their inherent drawbacks, such as substantial volume expansion upon melting, significant supercooling, and the propensity for phase separation, significantly limit the practical application of salts [147]. Metals and alloys [12]

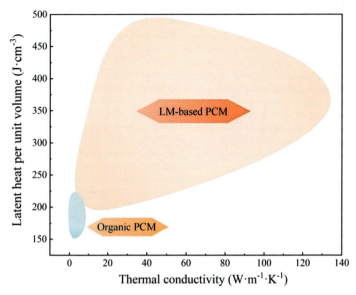

Fig. 18 Thermal conductivity and latent heat per unit volume of organic phase change material (PCM) and LM-based PCM. *Adapted from Ref. S. Wang, X. Zhao, J. Luo, L. Zhuang, D. Zou, Liquid metal (LM) and its composites in thermal management, Compos. Part A Appl. Sci. Manuf. 163 (2022) 107216, https://doi.org/10.1016/j.compositesa.2022.107216, with permission obtained from Elsevier [14].*

present promising prospects in latent heat energy storage systems due to their elevated energy storage density, minimal volume changes during melting, robust thermal stability, and thermal conductivity several times higher than other PCMs [149]. The existing range of thermal conductivity and latent heat per unit volume of organic PCM and LM-based PCM is shown in Fig. 18.

However, during practical usage, metal-based PCMs still face issues such as susceptibility to leakage, corrosion, and changes in volume. A promising solution to address these challenges is the microencapsulation of metal-based PCMs. This approach prevents the leakage of molten metal-based PCMs and safeguards against the loss of heat storage capacity resulting from PCM oxidation. In particular, microcapsules are widely recognized as an effective and advantageous solution. Microencapsulation involves the process of enveloping particles with coating materials or embedding them into a matrix to create microcapsules [147]. These microcapsules often feature an additional layer of air between the core and the shell, accommodating volumetric changes during the phase change [150]. The PCM enclosed within the core-shell structure is referred to as the core material, with the outer cladding being the shell material [151]. The size of the capsules typically ranges from a few nanometers to hundreds of microns. Microencapsulation resolves issues related to PCM leakage and corrosion and enhances stability and specific surface area [152]. The directions for encapsulating metal-based PCMs are depicted in Fig. 19.

When metal-based PCMs are encapsulated, they maintain their heat storage capabilities and gain mechanical properties, machinability, and a fixed shape. The resulting microcapsules can be combined with ceramics, silicone oil, and other materials to create solid and liquid heat storage PCMs, thus expanding the applications of metal-based PCM microcapsules. These applications span industrial waste heat recovery, solar photothermal power generation, heat transfer enhancement, and more [147].

Conversely, microencapsulation technology introduces specific challenges. As the shell materials are primarily organic or inorganic substances, the thermal conductivity of microcapsules may decrease to some extent compared to that of the metal-based core materials. Furthermore, due to the size effect and nucleation mode, metal-based microcapsules may exhibit higher supercooling compared to matrix metal-based PCMs. However,

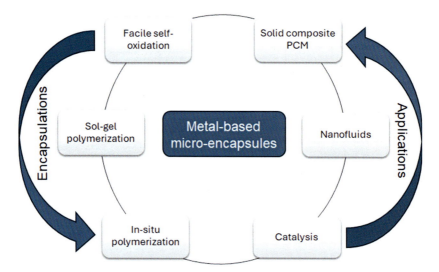

Fig. 19 Encapsulating methods and potential applications of metal-based phase-change material microcapsules. *Drawn based on Ref. S. Wang, K. Lei, Z. Wang, H. Wang, D. Zou, Metal-based phase change material (PCM) microcapsules/nanocapsules: fabrication, thermophysical characterization and application, Chem. Eng. J. 438 (2022) 135559, https://doi.org/10.1016/j.cej.2022.135559 [147].*

supercooling in microcapsules can be mitigated or reduced through shell modification, size control, the addition of nucleating agents, and external perturbations, among other techniques [147].

6. Our perspective and future directions

As we continue to explore and delve deeper into LM heat dissipation research and technology, it is crucial to carefully consider some scientific and technological challenges and future development directions. In the following, a summary of the perspective and future directions of GaLMs is presented:

i. Corrosion and compatibility: GaLMs exhibit significant corrosion effects on aluminum-based substrates, even at room temperature. While most other substrate materials, such as ceramics, copper, stainless steel, and more, are compatible with gallium- and bismuth-based LMs at lower operating temperatures, corrosion remains a concern as temperatures rise. To address this issue, innovative methods have been employed such as microencapsulation technology. This method has been utilized to create a barrier between the LM and the substrates, preventing direct contact and mitigating corrosion. Alternatively, techniques like coating and alloying can be employed to enhance the corrosion resistance of the substrates.

ii. Thermal conductivity: Pure GaLMs typically exhibit a thermal conductivity in the range of $7-35$ W·m^{-1}·K^{-1}. While they offer advantages over conventional thermal conductive materials like thermal silicone grease, there remains ample room for enhancement. One effective approach to maximize the thermal conductivity of LM-based composites involves incorporating nanoparticles with high thermal conductivity, such as nano-Cu, diamond, CNTs, and nano-graphene. Nonetheless, it is imperative to conduct further investigations into the dispersion stability of nanoparticles within LM.

iii. Low-temperature operation: Currently, Galinstan have a minimum melting point of approximately 11 °C, which means they freeze at lower temperatures. To meet the requirements of diverse application scenarios, it is necessary to explore the development of new GaLM variants with even lower melting points.

iv. Wettability: Oxidation enhances its ability to adhere to substrates, subsequently reducing contact thermal resistance and improving heat transfer efficiency. Nonetheless, the formation of oxide layers consumes LM materials, potentially diminishing their thermal properties. Therefore, there is a pressing need for advanced techniques that increase wettability and mitigate any compromise of thermal performance.

References

[1] S.M. Sohel Murshed, C.A. Nieto de Castro, A critical review of traditional and emerging techniques and fluids for electronics cooling, Renew. Sustain. Energy Rev. 78 (2017) 821–833, https://doi.org/10.1016/j.rser.2017.04.112

[2] X.M. Xu, R. He, Research on the heat dissipation performance of battery pack based on forced air cooling, J. Power Sources 240 (2013) 33–41, https://doi.org/10.1016/j.jpowsour.2013.03.004

[3] W. He, et al., Performance optimization of server water cooling system based on minimum energy consumption analysis, Appl. Energy 303 (2021) 117620, https://doi.org/10.1016/j.apenergy.2021.117620

[4] F. Bai, M. Chen, W. Song, Z. Feng, Y. Li, Y. Ding, Thermal management performances of PCM/water cooling-plate using for lithium-ion battery module based on non-uniform internal heat source, Appl. Therm. Eng. 126 (2017) 17–27, https://doi.org/10.1016/j.applthermaleng.2017.07.141

[5] Y. Zhang, J. Ma, N. Wei, J. Yang, Q.-X. Pei, Recent progress in the development of thermal interface materials: a review, Phys. Chem. Chem. Phys. 23 (2) (2021) 753–776, https://doi.org/10.1039/D0CP05514J

[6] A.L. Moore, L. Shi, Emerging challenges and materials for thermal management of electronics, Mater. Today 17 (4) (2014) 163–174, https://doi.org/10.1016/j.mattod.2014.04.003

[7] M. Lu, X. Zhang, J. Ji, X. Xu, Y. Zhang, Research progress on power battery cooling technology for electric vehicles, J. Energy Storage 27 (2020) 101155, https://doi.org/10.1016/j.est.2019.101155

[8] S. Moon, H. Kim, K. Lee, J. Park, Y. Kim, S.Q. Choi, 3D Printable concentrated liquid metal composite with high thermal conductivity, iScience 24 (10) (2021) 103183, https://doi.org/10.1016/j.isci.2021.103183

[9] N. Kazem, T. Hellebrekers, C. Majidi, Soft multifunctional composites and emulsions with liquid metals, Adv. Mater. 29 (27) (2017) 1605985, https://doi.org/10.1002/adma.201605985

[10] S. Chen, H.-Z. Wang, R.-Q. Zhao, W. Rao, J. Liu, Liquid metal composites, Matter 2 (6) (2020) 1446–1480, https://doi.org/10.1016/j.matt.2020.03.016

[11] R.S. Datta, et al., Flexible two-dimensional indium tin oxide fabricated using a liquid metal printing technique, Nat. Electron. 3 (1) (2020) 51–58, https://doi.org/10.1038/s41928-019-0353-8

[12] J. Yan, Y. Lu, G. Chen, M. Yang, Z. Gu, Advances in liquid metals for biomedical applications, Chem. Soc. Rev. 47 (8) (2018) 2518–2533, https://doi.org/10.1039/C7CS00309A

[13] J. Pacio, C. Singer, T. Wetzel, R. Uhlig, Thermodynamic evaluation of liquid metals as heat transfer fluids in concentrated solar power plants, Appl. Therm. Eng. 60 (1) (2013) 295–302, https://doi.org/10.1016/j.applthermaleng.2013.07.010

[14] S. Wang, X. Zhao, J. Luo, L. Zhuang, D. Zou, Liquid metal (LM) and its composites in thermal management, Compos. Part A Appl. Sci. Manuf. 163 (2022) 107216, https://doi.org/10.1016/j.compositesa.2022.107216

[15] H. Bark, P.S. Lee, Surface modification of liquid metal as an effective approach for deformable electronics and energy devices, Chem. Sci. 12 (8) (2021) 2760–2777, https://doi.org/10.1039/D0SC05310D

[16] Y. Lin, J. Genzer, M.D. Dickey, Attributes, fabrication, and applications of gallium-based liquid metal particles, Adv. Sci. 7 (12) (2020) 2000192, https://doi.org/10.1002/advs.202000192

[17] R.C. Chiechi, E.A. Weiss, M.D. Dickey, G.M. Whitesides, Eutectic Gallium–Indium (EGaIn): a moldable liquid metal for electrical characterization of self-assembled monolayers, Angew. Chemie Int. Ed. 47 (1) (2008) 142–144, https://doi.org/10.1002/anie.200703642

[18] M.D. Dickey, Stretchable and soft electronics using liquid metals, Adv. Mater. 29 (27) (2017) 1606425, https://doi.org/10.1002/adma.201606425

[19] S. Cheng, Z. Wu, Microfluidic electronics, Lab Chip 12 (16) (2012) 2782–2791, https://doi.org/10.1039/C2LC21176A

[20] S.-Y. Tang, et al., Liquid metal enabled pump, Proc. Natl. Acad. Sci. 111 (9) (2014) 3304–3309, https://doi.org/10.1073/pnas.1319878111

[21] J. Thelen, M.D. Dickey, T. Ward, A study of the production and reversible stability of EGaIn liquid metal microspheres using flow focusing, Lab Chip 12 (20) (2012) 3961–3967, https://doi.org/10.1039/C2LC40492C

[22] S.-Y. Tang, et al., Liquid-metal microdroplets formed dynamically with electrical control of size and rate, Adv. Mater. 28 (4) (2016) 604–609, https://doi.org/10.1002/adma.201503875

[23] B.L. Cumby, G.J. Hayes, M.D. Dickey, R.S. Justice, C.E. Tabor, J.C. Heikenfeld, Reconfigurable liquid metal circuits by Laplace pressure shaping, Appl. Phys. Lett. 101 (17) (2012) 174102, https://doi.org/10.1063/1.4764020

[24] V. Sivan, et al., Liquid metal marbles, Adv. Funct. Mater. 23 (2) (2013) 144–152, https://doi.org/10.1002/adfm.201200837

[25] F. Carle, K. Bai, J. Casara, K. Vanderlick, E. Brown, Development of magnetic liquid metal suspensions for magnetohydrodynamics, Phys. Rev. Fluids 2 (1) (2017) 13301, https://doi.org/10.1103/PhysRevFluids.2.013301

[26] I.A. De Castro, et al., A gallium-based magnetocaloric liquid metal ferrofluid, Nano Lett 17 (12) (2017) 7831–7838, https://doi.org/10.1021/acs.nanolett.7b04050

[27] M. Xiong, Y. Gao, J. Liu, Fabrication of magnetic nano liquid metal fluid through loading of Ni nanoparticles into gallium or its alloy, J. Magn. Magn. Mater. 354 (2014) 279–283, https://doi.org/10.1016/j.jmmm.2013.11.028

[28] M. Yu, X. Bian, T. Wang, J. Wang, Metal-based magnetic fluids with core–shell structure FeB@SiO2 amorphous particles, Soft Matter 13 (37) (2017) 6340–6348, https://doi.org/10.1039/C7SM01238A

[29] C. Yang, X. Bian, J. Qin, T. Guo, X. Zhao, Metal-based magnetic functional fluids with amorphous particles, RSC Adv 4 (103) (2014) 59541–59547, https://doi.org/10.1039/C4RA12481B

[30] L. Hu, H. Wang, X. Wang, X. Liu, J. Guo, J. Liu, Magnetic liquid metals manipulated in the three-dimensional free space, ACS Appl. Mater. Interfaces 11 (8) (2019) 8685–8692, https://doi.org/10.1021/acsami.8b22699

[31] P. Aussillous, D. Quéré, Liquid marbles, Nature 411 (6840) (2001) 924–927, https://doi.org/10.1038/35082026

[32] W. Babatain, M.S. Kim, M.M. Hussain, From droplets to devices: Recent advances in liquid metal droplet enabled electronics, Adv. Funct. Mater. (2023) 2308116, https://doi.org/10.1002/adfm.202308116

[33] V.A. Beck, et al., A combined numerical and experimental study to elucidate primary breakup dynamics in liquid metal droplet-on-demand printing, Phys. Fluids 32 (11) (2020) 112020, https://doi.org/10.1063/5.0029438

[34] C. Ladd, J.-H. So, J. Muth, M.D. Dickey, 3D printing of free standing liquid metal microstructures, Adv. Mater. 25 (36) (2013) 5081–5085, https://doi.org/10.1002/adma.201301400

[35] M.G. Mohammed, A. Xenakis, M.D. Dickey, Production of liquid metal spheres by molding, Metals 4 (4) (2014) 465–476, https://doi.org/10.3390/met4040465

[36] L. Tian, M. Gao, L. Gui, A microfluidic chip for liquid metal droplet generation and sorting, Micromachines 8 (2) (2017), https://doi.org/10.3390/mi8020039

[37] K. Akyildiz, J.-H. Kim, J.-H. So, H.-J. Koo, Recent progress on micro- and nanoparticles of gallium-based liquid metals: from preparation to applications, J. Ind. Eng. Chem. 116 (2022) 120–141, https://doi.org/10.1016/j.jiec.2022.09.046

[38] B. Gol, et al., Continuous transfer of liquid metal droplets across a fluid–fluid interface within an integrated microfluidic chip, Lab Chip 15 (11) (2015) 2476–2485, https://doi.org/10.1039/C5LC00415B

[39] I.D. Tevis, L.B. Newcomb, M. Thuo, Synthesis of liquid core–shell particles and solid patchy multicomponent particles by shearing liquids into complex particles (SLICE), Langmuir 30 (47) (2014) 14308–14313, https://doi.org/10.1021/la5035118

[40] J. Thelen, M.D. Dickey, T. Ward, A study of the production and reversible stability of EGaIn liquid metal microspheres using flow focusing, Lab Chip 12 (20) (2012) 3961–3967, https://doi.org/10.1039/C2LC40492C

[41] C.N. Baroud, F. Gallaire, R. Dangla, Dynamics of microfluidic droplets, Lab Chip 10 (16) (2010) 2032–2045, https://doi.org/10.1039/C001191F

[42] B. Gol, M.E. Kurdzinski, F.J. Tovar-Lopez, P. Petersen, A. Mitchell, K. Khoshmanesh, Hydrodynamic directional control of liquid metal droplets within a microfluidic flow focusing system, Appl. Phys. Lett. 108 (16) (2016) 164101, https://doi.org/10.1063/1.4947272

[43] T. Hutter, W.-A.C. Bauer, S.R. Elliott, W.T.S. Huck, Formation of spherical and non-spherical eutectic gallium-indium liquid-metal microdroplets in microfluidic channels at room temperature, Adv. Funct. Mater. 22 (12) (2012) 2624–2631, https://doi.org/10.1002/adfm.201200324

[44] A. Yamaguchi, Y. Mashima, T. Iyoda, Reversible size control of liquid-metal nanoparticles under ultrasonication, Angew. Chem. Int. Ed. 54 (43) (2015) 12809–12813, https://doi.org/10.1002/anie.201506469

[45] V.B. Kumar, A. Gedanken, Z. Porat, Sonochemistry of molten gallium, Ultrason. Sonochem. 95 (2023) 106364, https://doi.org/10.1016/j.ultsonch.2023.106364

[46] J. Jeong, J.-B. Lee, S.K. Chung, D. Kim, Electromagnetic three dimensional liquid metal manipulation, Lab Chip 19 (19) (2019) 3261–3267, https://doi.org/10.1039/C9LC00503J

[47] L. Majidi, D. Gritsenko, J. Xu, Gallium-Based room-temperature liquid metals: actuation and manipulation of droplets and flows, Front. Mech. Eng. 3 (2017) ([Online]. Available), ⟨https://www.frontiersin.org/articles/10.3389/fmech.2017.00009⟩.

[48] D. Kim, J.-B. Lee, Magnetic-field-induced liquid metal droplet manipulation, J. Korean Phys. Soc. 66 (2) (2015) 282–286, https://doi.org/10.3938/jkps.66.282

[49] I.D. Joshipura, H.R. Ayers, C. Majidi, M.D. Dickey, Methods to pattern liquid metals, J. Mater. Chem. C 3 (16) (2015) 3834–3841, https://doi.org/10.1039/C5TC00330J

[50] T. Daeneke, et al., Liquid metals: fundamentals and applications in chemistry, Chem. Soc. Rev. 47 (11) (2018) 4073–4111, https://doi.org/10.1039/C7CS00043J

[51] K. Khoshmanesh, et al., Liquid metal enabled microfluidics, Lab Chip 17 (6) (2017) 974–993, https://doi.org/10.1039/C7LC00046D

[52] M. Duan, X. Zhu, X. Shan, H. Wang, S. Chen, J. Liu, Responsive liquid metal droplets: from bulk to nano, Nanomaterials 12 (8) (2022), https://doi.org/10.3390/nano12081289

[53] B. Yuan, S. Tan, Y. Zhou, J. Liu, Self-powered macroscopic Brownian motion of spontaneously running liquid metal motors, Sci. Bull. 60 (13) (2015) 1203–1210, https://doi.org/10.1007/s11434-015-0836-6

[54] L. Ren, X. Xu, Y. Du, K. Kalantar-Zadeh, S.X. Dou, Liquid metals and their hybrids as stimulus–responsive smart materials, Mater. Today 34 (2020) 92–114, https://doi.org/10.1016/j.mattod.2019.10.007

[55] X. Wang, R. Guo, J. Liu, Liquid metal based soft robotics: materials, designs, and applications, Adv. Mater. Technol. 4 (2) (2019) 1800549, https://doi.org/10.1002/admt.201800549

[56] D. Kim, et al., Recovery of nonwetting characteristics by surface modification of gallium-based liquid metal droplets using hydrochloric acid vapor, ACS Appl. Mater. Interfaces 5 (1) (2013) 179–185, https://doi.org/10.1021/am302357t

[57] F. Mugele, J.-C. Baret, Electrowetting: from basics to applications, J. Phys. Condens. Matter 17 (28) (2005) R705, https://doi.org/10.1088/0953-8984/17/28/R01

[58] D.C. Grahame, The electrical double layer and the theory of electrocapillarity, Chem. Rev. 41 (3) (1947) 441–501, https://doi.org/10.1021/cr60130a002

[59] A. Frumkin, N. Polianovskaya, N. Grigoryev, I. Bagotskaya, Electrocapillary phenomena on gallium, Electrochim. Acta 10 (8) (1965) 793–802, https://doi.org/10.1016/0013-4686(65)80044-5

[60] V.A. Marichev, Maximum surface tension and optimum surface electron density, Colloids Surf. A Physicochem. Eng. Asp. 389 (1) (2011) 63–68, https://doi.org/10.1016/j.colsurfa.2011.08.050

[61] I.A. Bagotskaya, A.M. Morozov, N.B. Grigoryev, On the zero-charge potential of gallium in aqueous solutions and the origin of high capacitance at the gallium/solution interface, Electrochim. Acta 13 (4) (1968) 873–879, https://doi.org/10.1016/0013-4686(68)85020-0

[62] J.L. Jackel, S. Hackwood, G. Beni, Electrowetting optical switch, Appl. Phys. Lett. 40 (1) (1982) 4–5, https://doi.org/10.1063/1.92920

[63] G. Beni, S. Hackwood, J.L. Jackel, Continuous electrowetting effect, Appl. Phys. Lett. 40 (10) (1982) 912–914, https://doi.org/10.1063/1.92952

[64] C.B. Eaker, M.D. Dickey, Liquid metal actuation by electrical control of interfacial tension, Appl. Phys. Rev. 3 (3) (2016) 31103, https://doi.org/10.1063/1.4959898

[65] A.A. Kornyshev, A.R. Kucernak, M. Marinescu, C.W. Monroe, A.E.S. Sleightholme, M. Urbakh, Ultra-low-voltage electrowetting, J. Phys. Chem. C 114 (35) (2010) 14885–14890, https://doi.org/10.1021/jp101051e

[66] T. Young III, An essay on the cohesion of fluids, Philos. Trans. R. Soc. London 95 (1997) 65–87, https://doi.org/10.1098/rstl.1805.0005

[67] D.A. Antelmi, J.N. Connor, R.G. Horn, Electrowetting measurements with mercury showing mercury/mica interfacial energy depends on charging, J. Phys. Chem. B 108 (3) (2004) 1030–1037, https://doi.org/10.1021/jp036371u

[68] S. Arscott, Electrowetting at a liquid metal-oxide-semiconductor junction, Appl. Phys. Lett. 103 (14) (2013) 144101, https://doi.org/10.1063/1.4822308

[69] Z. Wan, H. Zeng, A. Feinerman, Reversible electrowetting of liquid-metal droplet, J. Fluids Eng. 129 (4) (2006) 388–394, https://doi.org/10.1115/1.2436582

[70] Z. Wan, H. Zeng, A. Feinerman, Reversible electrowetting of liquid-metal droplet, J. Fluids Eng. 129 (4) (2006) 388–394, https://doi.org/10.1115/1.2436582

[71] T. Liu, P. Sen, C.-J. Kim, Characterization of nontoxic liquid-metal alloy galinstan for applications in microdevices, J. Microelectromech. Syst. 21 (2) (2012) 443–450, https://doi.org/10.1109/JMEMS.2011.2174421

[72] X. Deng, Z. Shao, Y. Zhao, Solutions to the drawbacks of photothermal and photodynamic cancer therapy, Adv. Sci. 8 (3) (2021) 2002504, https://doi.org/10.1002/advs.202002504

[73] J. Shu, S.-Y. Tang, Z. Feng, W. Li, X. Li, S. Zhang, Unconventional locomotion of liquid metal droplets driven by magnetic fields, Soft Matter 14 (35) (2018) 7113–7118, https://doi.org/10.1039/C8SM01281D

[74] L. Wang, J. Liu, Electromagnetic rotation of a liquid metal sphere or pool within a solution, Proc. R. Soc. A Math. Phys. Eng. Sci. 471 (2178) (2015) 20150177, https://doi.org/10.1098/rspa.2015.0177

[75] Y. Yu, E. Miyako, Alternating-magnetic-field-mediated wireless manipulations of a liquid metal for therapeutic bioengineering, iScience 3 (2018) 134–148, https://doi.org/10.1016/j.isci.2018.04.012

[76] P. Aussillous, D. Quéré, Properties of liquid marbles, Proc. R. Soc. A Math. Phys. Eng. Sci. 462 (2067) (2006) 973–999, https://doi.org/10.1098/rspa.2005.1581

[77] D. Kim, J.-B. Lee, Magnetic-field-induced liquid metal droplet manipulation, J. Korean Phys. Soc. 66 (2) (2015) 282–286, https://doi.org/10.3938/jkps.66.282

[78] J.M. McMahon, G.C. Schatz, S.K. Gray, Plasmonics in the ultraviolet with the poor metals Al, Ga, In, Sn, Tl, Pb, and Bi, Phys. Chem. Chem. Phys. 15 (15) (2013) 5415–5423, https://doi.org/10.1039/C3CP43856B

[79] P. Albella, et al., Shape matters: plasmonic nanoparticle shape enhances interaction with dielectric substrate, Nano Lett. 11 (9) (2011) 3531–3537, https://doi.org/10.1021/nl201783v

[80] Y. Lu, et al., Enhanced endosomal escape by light-fueled liquid-metal transformer, Nano Lett. 17 (4) (2017) 2138–2145, https://doi.org/10.1021/acs.nanolett.6b04346

[81] X. Tang, et al., Photochemically induced motion of liquid metal marbles, Appl. Phys. Lett. 103 (17) (2013) 174104, https://doi.org/10.1063/1.4826923

[82] X. Wang, et al., Soft and moldable Mg-doped liquid metal for conformable skin tumor photothermal therapy, Adv. Healthc. Mater. 7 (14) (2018) 1800318, https://doi.org/10.1002/adhm.201800318

[83] J. Jeon, J.B. Lee, S.K. Chung, D. Kim, Acoustic wave-driven oxide dependant dynamic behavior of liquid metal droplet for inkjet applications, Mater. Res. Express 10 (5) (2023) 55701, https://doi.org/10.1088/2053-1591/accf62

[84] D. Wang, C. Gao, C. Zhou, Z. Lin, Q. He, Leukocyte membrane-coated liquid metal nanoswimmers for actively targeted delivery and synergistic chemophotothermal therapy, Research 2020 (2023), https://doi.org/10.34133/2020/3676954

[85] Y. Gao, Y. Bando, Carbon nanothermometer containing gallium, Nature 415 (6872) (2002) 599, https://doi.org/10.1038/415599a

[86] S. Chen, et al., Generalized way to make temperature tunable conductor–insulator transition liquid metal composites in a diverse range, Mater. Horizons 6 (9) (2019) 1854–1861, https://doi.org/10.1039/C9MH00650H

[87] X. Sun, et al., Low-temperature triggered shape transformation of liquid metal microdroplets, ACS Appl. Mater. Interfaces 12 (34) (2020) 38386–38396, https://doi.org/10.1021/acsami.0c10409

[88] R.A. Bilodeau, D.Y. Zemlyanov, R.K. Kramer, Liquid metal switches for environmentally responsive electronics, Adv. Mater. Interfaces 4 (5) (2017) 1600913, https://doi.org/10.1002/admi.201600913

[89] S. Chen, L. Wang, Q. Zhang, J. Liu, Liquid metal fractals induced by synergistic oxidation, Sci. Bull. 63 (22) (2018) 1513–1520, https://doi.org/10.1016/j.scib.2018.10.008

[90] S. Chen, X. Yang, Y. Cui, J. Liu, Self-growing and serpentine locomotion of liquid metal induced by copper ions, ACS Appl. Mater. Interfaces 10 (27) (2018) 22889–22895, https://doi.org/10.1021/acsami.8b07649

[91] S. Handschuh-Wang, Y. Chen, L. Zhu, X. Zhou, Analysis and transformations of room-temperature liquid metal interfaces—a closer look through interfacial tension, ChemPhysChem 19 (13) (2018) 1584–1592, https://doi.org/10.1002/cphc.201800129

[92] L. Hu, J. Li, J. Tang, J. Liu, Surface effects of liquid metal amoeba, Sci. Bull. 62 (10) (2017) 700–706, https://doi.org/10.1016/j.scib.2017.04.015

[93] L. Wang, J. Liu, Graphite induced periodical self-actuation of liquid metal, RSC Adv. 6 (65) (2016) 60729–60735, https://doi.org/10.1039/C6RA12177B

[94] R.C. Gough, et al., Self-actuation of liquid metal via redox reaction, ACS Appl. Mater. Interfaces 8 (1) (2016) 6–10, https://doi.org/10.1021/acsami.5b09466

[95] J. Zhang, Y. Yao, L. Sheng, J. Liu, Self-fueled biomimetic liquid metal mollusk, Adv. Mater. 27 (16) (2015) 2648–2655, https://doi.org/10.1002/adma.201405438

[96] L. Sheng, Z. He, Y. Yao, J. Liu, Transient state machine enabled from the colliding and coalescence of a swarm of autonomously running liquid metal motors, Small 11 (39) (2015) 5253–5261, https://doi.org/10.1002/smll.201501364

[97] J. Zhang, Y. Yao, J. Liu, Autonomous convergence and divergence of the self-powered soft liquid metal vehicles, Sci. Bull. 60 (10) (2015) 943–951, https://doi.org/10.1007/s11434-015-0786-z

[98] F. Li, et al., Magnetically- and electrically-controllable functional liquid metal droplets, Adv. Mater. Technol. 4 (3) (2019) 1800694, https://doi.org/10.1002/admt.201800694

[99] C. Du, X. Wu, M. He, Y. Zhang, R. Zhang, C.-M. Dong, Polymeric photothermal agents for cancer therapy: recent progress and clinical potential, J. Mater. Chem. B 9 (6) (2021) 1478–1490, https://doi.org/10.1039/D0TB02659J

[100] A. Hak, V. Ravasaheb Shinde, A.K. Rengan, A review of advanced nanoformulations in phototherapy for cancer therapeutics, Photodiagnosis Photodyn. Ther. 33 (2021) 102205, https://doi.org/10.1016/j.pdpdt.2021.102205

[101] Y. Hou, P. Zhang, D. Wang, J. Liu, W. Rao, Liquid metal hybrid platform-mediated ice–fire dual noninvasive conformable melanoma therapy, ACS Appl. Mater. Interfaces 12 (25) (2020) 27984–27993, https://doi.org/10.1021/acsami.0c06023

[102] A. Chicheł, J. Skowronek, M. Kubaszewska, M. Kanikowski, Hyperthermia—description of a method and a review of clinical applications, Rep. Pract. Oncol. Radiother. 12 (5) (2007) 267–275, https://doi.org/10.1016/S1507-1367(10)60065-X

[103] J.-J. Hu, et al., Photo-controlled liquid metal nanoparticle-enzyme for starvation/photothermal therapy of tumor by win-win cooperation, Biomaterials 217 (2019) 119303, https://doi.org/10.1016/j.biomaterials.2019.119303

[104] K. Yamagishi, W. Zhou, T. Ching, S.Y. Huang, M. Hashimoto, Ultra-deformable and tissue-adhesive liquid metal antennas with high wireless powering efficiency, Adv. Mater. 33 (26) (2021) 2008062, https://doi.org/10.1002/adma.202008062

[105] J. Kadkhoda, A. Tarighatnia, J. Barar, A. Aghanejad, S. Davaran, Recent advances and trends in nanoparticles based photothermal and photodynamic therapy, Photodiagnosis Photodyn. Ther. 37 (2022) 102697, https://doi.org/10.1016/j.pdpdt.2021.102697

[106] W. Lee, C.E. Lee, H.J. Kim, K. Kim, Current progress in gallium-based liquid metals for combinatory phototherapeutic anticancer applications, Colloids Surf. B Biointerfaces 226 (2023) 113294, https://doi.org/10.1016/j.colsurfb.2023.113294

[107] J. Yan, et al., Shape-controlled synthesis of liquid metal nanodroplets for photothermal therapy, Nano Res. 12 (6) (2019) 1313–1320, https://doi.org/10.1007/s12274-018-2262-y

[108] N. Xia, et al., Multifunctional and flexible ZrO2-coated EGaIn nanoparticles for photothermal therapy, Nanoscale 11 (21) (2019) 10183–10189, https://doi.org/10.1039/C9NR01963D

[109] P. Zhu, et al., Inorganic nanoshell-stabilized liquid metal for targeted photonanomedicine in NIR-II biowindow, Nano Lett. 19 (3) (2019) 2128–2137, https://doi.org/10.1021/acs.nanolett.9b00364

[110] X. Sun, et al., Shape tunable gallium nanorods mediated tumor enhanced ablation through near-infrared photothermal therapy, Nanoscale 11 (6) (2019) 2655–2667, https://doi.org/10.1039/C8NR08296K

[111] S.A. Chechetka, Y. Yu, X. Zhen, M. Pramanik, K. Pu, E. Miyako, Light-driven liquid metal nanotransformers for biomedical theranostics, Nat. Commun. 8 (1) (2017) 15432, https://doi.org/10.1038/ncomms15432

[112] J.-J. Hu, et al., Immobilized liquid metal nanoparticles with improved stability and photothermal performance for combinational therapy of tumor, Biomaterials 207 (2019) 76–88, https://doi.org/10.1016/j.biomaterials.2019.03.043

[113] Y. Zhang, et al., Synthesis of liquid gallium@reduced graphene oxide core–shell nanoparticles with enhanced photoacoustic and photothermal performance, J. Am. Chem. Soc. 144 (15) (2022) 6779–6790, https://doi.org/10.1021/jacs.2c00162

[114] Z. Guo, et al., Galvanic replacement reaction for in situ fabrication of litchi-shaped heterogeneous liquid metal-Au nano-composite for radio-photothermal cancer therapy, Bioact. Mater. 6 (3) (2021) 602–612, https://doi.org/10.1016/j.bioactmat.2020.08.033

[115] J. Oh, et al., Jumping-droplet electronics hot-spot cooling, Appl. Phys. Lett. 110 (12) (2017) 123107, https://doi.org/10.1063/1.4979034

[116] R. Kandasamy, X.-Q. Wang, A.S. Mujumdar, Transient cooling of electronics using phase change material (PCM)-based heat sinks, Appl. Therm. Eng. 28 (8) (2008) 1047–1057, https://doi.org/10.1016/j.applthermaleng.2007.06.010

[117] J.B. Boreyko, Y. Zhao, C.-H. Chen, Planar jumping-drop thermal diodes, Appl. Phys. Lett. 99 (23) (2011) 234105, https://doi.org/10.1063/1.3666818

[118] N. Li, J. Ren, L. Wang, G. Zhang, P. Hänggi, B. Li, Colloquium: phononics: manipulating heat flow with electronic analogs and beyond, Rev. Mod. Phys. 84 (3) (2012) 1045–1066, https://doi.org/10.1103/RevModPhys.84.1045

[119] T. Yang, et al., Millimeter-scale liquid metal droplet thermal switch, Appl. Phys. Lett. 112 (6) (2018) 63505, https://doi.org/10.1063/1.5013623

[120] T. Yang, et al., An integrated liquid metal thermal switch for active thermal management of electronics, IEEE Trans. Compon. Packag. Manuf. Technol. 9 (12) (2019) 2341–2351, https://doi.org/10.1109/TCPMT.2019.2930089

[121] A. Priimagi, C.J. Barrett, A. Shishido, Recent twists in photoactuation and photoalignment control, J. Mater. Chem. C 2 (35) (2014) 7155–7162, https://doi.org/10.1039/C4TC01236D

[122] Y. Hu, Z. Li, T. Lan, W. Chen, Photoactuators for direct optical-to-mechanical energy conversion: from nanocomponent assembly to macroscopic deformation, Adv. Mater. 28 (47) (2016) 10548–10556, https://doi.org/10.1002/adma.201602685

[123] L. Xu, Q. Peng, X. Zhao, P. Li, J. Xu, X. He, A photoactuator based on stiffness-variable carbon nanotube nanocomposite yarn, ACS Appl. Mater. Interfaces 12 (36) (2020) 40711–40718, https://doi.org/10.1021/acsami.0c14222

[124] H. Arazoe, et al., An autonomous actuator driven by fluctuations in ambient humidity, Nat. Mater. 15 (10) (2016) 1084–1089, https://doi.org/10.1038/nmat4693

[125] V.K. PamulaK. Chakrabarty, Cooling of integrated circuits using droplet-based microfluidics, in: Proceedings of the 13th ACM Great Lakes Symposium on VLSI, in GLSVLSI '03. Association for Computing Machinery, New York, 2003, pp. 84–87. doi: 10.1145/764808.764831.

[126] X. Li, et al., Controlled transformation of liquid metal microspheres in aqueous solution triggered by growth of GaOOH, ACS Omega 7 (9) (2022) 7912–7919, https://doi.org/10.1021/acsomega.1c06897

[127] P. Bhuyan, et al., Soft and stretchable liquid metal composites with shape memory and healable conductivity, ACS Appl. Mater. Interfaces 13 (24) (2021) 28916–28924, https://doi.org/10.1021/acsami.1c06786

[128] M.D. Bartlett, et al., High thermal conductivity in soft elastomers with elongated liquid metal inclusions, Proc. Natl. Acad. Sci. 114 (9) (2017) 2143–2148, https://doi.org/10.1073/pnas.1616377114

[129] G. Toci, et al., Gold nanostars embedded in PDMS films: a photothermal material for antibacterial applications, Nanomaterials 11 (12) (2021), https://doi.org/10.3390/nano11123252

[130] P. Grisoli, et al., PVA films with mixed silver nanoparticles and gold nanostars for intrinsic and photothermal antibacterial action, Nanomaterials 11 (6) (2021), https://doi.org/10.3390/nano11061387

[131] T. Gan, W. Shang, S. Handschuh-Wang, X. Zhou, Light-induced shape morphing of liquid metal nanodroplets enabled by polydopamine coating, Small 15 (9) (2019) 1804838, https://doi.org/10.1002/smll.201804838

[132] K.-T. Lin, H. Lin, T. Yang, B. Jia, Structured graphene metamaterial selective absorbers for high efficiency and omnidirectional solar thermal energy conversion, Nat. Commun. 11 (1) (2020) 1389, https://doi.org/10.1038/s41467-020-15116-z

[133] G. Liu, J. Xu, K. Wang, Solar water evaporation by black photothermal sheets, Nano Energy 41 (2017) 269–284, https://doi.org/10.1016/j.nanoen.2017.09.005

[134] S. Yang, Y. Zhang, J. Bai, Y. He, X. Zhao, J. Zhang, Integrating dual-interfacial liquid metal based nanodroplet architectures and micro-nanostructured engineering for high efficiency solar energy harvesting, ACS Nano 16 (9) (2022) 15086–15099, https://doi.org/10.1021/acsnano.2c06245

[135] Z. Yan, M. Jin, Z. Li, G. Zhou, L. Shui, Droplet-based microfluidic thermal management methods for high performance electronic devices, Micromachines 10 (2) (2019), https://doi.org/10.3390/mi10020089

[136] K. Mohseni, Effective cooling of integrated circuits using liquid alloy electrowetting, in: Semiconductor Thermal Measurement and Management IEEE Twenty First Annual IEEE Symposium, 2005, pp. 20–25. https://doi.org/10.1109/STHERM.2005.1412154.

[137] E. Baird, K. Mohseni, Digitized heat transfer: a new paradigm for thermal management of compact micro systems, IEEE Trans. Compon. Packag. Technol. 31 (1) (2008) 143–151, https://doi.org/10.1109/TCAPT.2008.916810

[138] J.Y. Zhu, S.-Y. Tang, K. Khoshmanesh, K. Ghorbani, An integrated liquid cooling system based on galinstan liquid metal droplets, ACS Appl. Mater. Interfaces 8 (3) (2016) 2173–2180, https://doi.org/10.1021/acsami.5b10769

[139] G.G. Guymon, M.H. Malakooti, Multifunctional liquid metal polymer composites, J. Polym. Sci. 60 (8) (2022) 1300–1327, https://doi.org/10.1002/pol.20210867

[140] Y. Ji, H. Yan, X. Xiao, J. Xu, Y. Li, C. Chang, Excellent thermal performance of gallium–based liquid metal alloy as thermal interface material between aluminum substrates, Appl. Therm. Eng. 166 (2020) 114649, https://doi.org/10.1016/j.applthermaleng.2019.114649

[141] D.D.L. Chung, Thermal interface materials, J. Mater. Eng. Perform. 10 (1) (2001) 56–59, https://doi.org/10.1361/105994901770345358

[142] S. Ki, et al., Gallium-based liquid metal alloy incorporating oxide-free copper nanoparticle clusters for high-performance thermal interface materials, Int. J. Heat Mass Transf. 170 (2021) 121012, https://doi.org/10.1016/j.ijheatmasstransfer.2021.121012

[143] H. Wang, et al., A highly stretchable liquid metal polymer as reversible transitional insulator and conductor, Adv. Mater. 31 (23) (2019) 1901337, https://doi.org/10.1002/adma.201901337

[144] A.B.M.T. Haque, R. Tutika, R.L. Byrum, M.D. Bartlett, Programmable liquid metal microstructures for multifunctional soft thermal composites, Adv. Funct. Mater. 30 (25) (2020) 2000832, https://doi.org/10.1002/adfm.202000832

[145] R.K. Sharma, P. Ganesan, V.V. Tyagi, H.S.C. Metselaar, S.C. Sandaran, Developments in organic solid–liquid phase change materials and their applications in thermal energy storage, Energy Convers. Manag. 95 (2015) 193–228, https://doi.org/10.1016/j.enconman.2015.01.084

[146] A. Sharma, V.V. Tyagi, C.R. Chen, D. Buddhi, Review on thermal energy storage with phase change materials and applications, Renew. Sustain. Energy Rev. 13 (2) (2009) 318–345, https://doi.org/10.1016/j.rser.2007.10.005

[147] S. Wang, K. Lei, Z. Wang, H. Wang, D. Zou, Metal-based phase change material (PCM) microcapsules/nanocapsules: fabrication, thermophysical characterization and application, Chem. Eng. J. 438 (2022) 135559, https://doi.org/10.1016/j.cej.2022.135559

[148] Y. Lin, G. Alva, G. Fang, Review on thermal performances and applications of thermal energy storage systems with inorganic phase change materials, Energy 165 (2018) 685–708, https://doi.org/10.1016/j.energy.2018.09.128

[149] P.J. Shamberger, N.M. Bruno, Review of metallic phase change materials for high heat flux transient thermal management applications, Appl. Energy 258 (2020) 113955, https://doi.org/10.1016/j.apenergy.2019.113955

[150] H. Nazir, et al., Recent developments in phase change materials for energy storage applications: a review, Int. J. Heat Mass Transf. 129 (2019) 491–523, https://doi.org/10.1016/j.ijheatmasstransfer.2018.09.126

[151] B. Xu, R.A. Ghossein, The contribution of molecular pathology to the classification of thyroid tumors, Diagn. Histopathol. 24 (3) (2018) 87–94, https://doi.org/10.1016/j.mpdhp.2018.02.001

[152] Z. Chen, G. Fang, Preparation and heat transfer characteristics of microencapsulated phase change material slurry: a review, Renew. Sustain. Energy Rev. 15 (9) (2011) 4624–4632, https://doi.org/10.1016/j.rser.2011.07.090

CHAPTER FIVE

Thermal transport in engineered cellular materials: A contemporary perspective

Prashant Singh[a] and Roop L. Mahajan[b,*]

[a]Department of Mechanical, Aerospace & Biomedical Engineering, University of Tennessee, Knoxville, TN, United States
[b]Department of Mechanical Engineering, Virginia Tech, Blacksburg, VA, United States
*Corresponding author. e-mail address: mahajanr@vt.edu

Contents

1. Introduction	184
2. Stochastic foams	185
2.1 Early investigations	185
2.2 Looking ahead	194
3. Pore-scale flow and thermal transport in stochastic cellular materials	195
4. Pore-scale flow and thermal transport in architectured cellular materials: strut-based	205
4.1 Forced convection with air as working fluid	207
4.2 Forced convection with some other working fluids (water, hydrocarbons, particles, supercritical carbon dioxide)	221
4.3 Architectured cellular materials for applications requiring conformal cooling solutions	226
5. Concluding remarks	227
References	229

Abstract

High porosity metal foams have been widely adopted in enhanced heat transfer applications, due to their ability to dissipate large heat flux levels. They offer high surface area-to-volume ratio, promote flow mixing and thermal dispersion due to flow tortuosity. The field of porous media research has experienced a transformative shift, propelled by advancements in manufacturing techniques, non-intrusive diagnostics, and computational capabilities. This shift has led to a widespread exploration of architectured cellular materials, leveraging the manufacturing-enabled design freedom. This article specifically focuses on recent developments in this area, with an emphasis on the pore- and strut-level flow and thermal transport for single-phase flows involving various working fluids such as air, water, hydrocarbons, supercritical carbon dioxide, and particles. The multifunctional attributes of recently developed engineered cellular materials and their wide range of applications are discussed. The article concludes by outlining the future directions in engineered cellular materials.

Advances in Heat Transfer, Volume 57
ISSN 0065-2717, https://doi.org/10.1016/bs.aiht.2024.02.001
Copyright © 2024 Elsevier Inc. All rights are reserved, including those for text and data mining, AI training, and similar technologies

Nomenclature

a_{sf}	surface area-to-volume ratio.
A	area.
C_f	inertial coefficient.
d_f	fiber/strut diameter.
d_p	pore diameter.
G	shape function.
h_{sf}	interstitial heat transfer coefficient.
k	thermal conductivity.
k_{eff}	effective thermal conductivity.
K	permeability.
L	length.
Nu	Nusselt number.
p	pressure.
Δp	pressure drop.
q	heat transfer rate.
Re_{d_f}	fiber diameter-based Reynolds number.
Re_K	permeability-based Reynolds number.
R_{th}	thermal resistance.
S_h	source term.
T_f	fluid-phase temperature.
T_s	solid-phase temperature.
$\langle u \rangle$	velocity in porous zone.
v	velocity.
$\dot{W}$	pumping power.
x	direction.

Greek symbols

ε	porosity.
μ	dynamic viscosity.
ρ	density.
θ_f	normalized temperature of fluid phase.

1. Introduction

Cellular materials consist of solid matrix and void spaces (pores) which can be either interconnected or isolated from each other. The constituent representative unit cells of the cellular materials can be stochastic or ordered in nature, depending upon the manufacturing process. These two broad categories represent two different approaches in the design and fabrication of cellular materials. Stochastic foams comprise of randomly organized unit cells—an inherent outcome of the manufacturing process deployed.

In contrast, foams manufactured with ordered arrangement of unit cells have a deliberately designed cellular structure, allowing control over cell size, shape, distribution, and connectivity. This precision enables tailoring the material for multifunctional attributes, resulting in more predictable and reproducible performance compared to stochastic foams. The choice between the two depends on the specific application requirements, including mechanical properties, weight considerations, and cost factors.

The response of cellular materials, as illustrated in Fig. 1, to thermal, mechanical, electrical, and acoustical stimuli depends on the solid matrix, the strut (also referred as "fibers" in metal foam literature) interconnections (topology), the porosity, and the properties of the media occupying the void space. These materials find applications in diverse fields, serving as lightweight heat exchangers [2–5], mechanical energy absorbers [6–9], noise reducers [10–12], biomedical implants [13,14]. The distinctive thermal transport and mechanical properties of cellular materials stem from their high specific surface area (m^2/m^3), tortuous flow paths, and lightweight nature resulting from high porosity. From a thermal transport perspective, metallic cellular materials offer superior effective thermal conductivity and a high interstitial convective heat transfer coefficient, while the pores allow the accommodation of a wide range of thermal transport agents, driven by density - and/or pressure gradients, making them truly multifunctional. The effectiveness of cellular materials in enhancing heat transfer has been well-established across a myriad of applications in diverse industrial and scientific domains. Within the realm of stochastic cellular materials, an extensive body of published literature delves into various aspects of thermal transport. For comprehensive review of flow and thermal transport properties of stochastic foams, refer to [15,16]. In this chapter, we have taken a chronological approach to discuss the evolution of cellular materials and their engineering. We explore how research has progressed over the years, propelled by advancements in non-intrusive diagnostics, computational power, and innovations in manufacturing.

2. Stochastic foams
2.1 Early investigations

The early investigations on flow and thermal transport primarily concentrated on stochastic metal foams manufactured via the foaming process - a technique involving the passage of an inert gas through a molten metal pool. The gas bubbles rise through the molten pool, leaving behind an

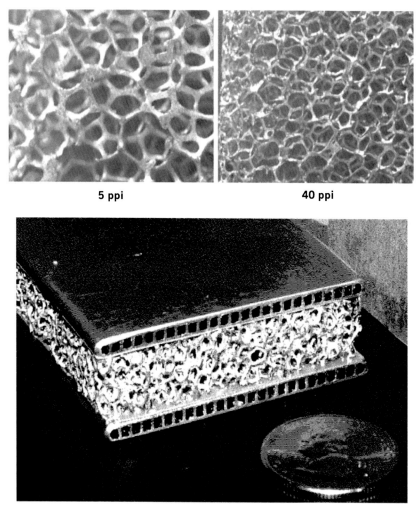

Fig. 1 (Top) High porosity aluminum foams through the foaming process [1], an integrated heat exchanger involving metal foam and minichannels [2].

interconnected network of struts after solidification. This manufacturing approach is based on the principle of minimum free surface energy dictating the equilibrium state. The resulting representative unit cell closely approximates the tetrakaidecahedron (TKD) shape, especially for high porosity metal foams. Typically, these metal foams exhibit pore densities ranging from 5 to 40 pores per inch and porosities exceeding 0.9 (Fig. 1). The creation of the TKD unit cell geometry can take various forms, such as body-centered and sphere-centered, as shown in Fig. 2. It is important to

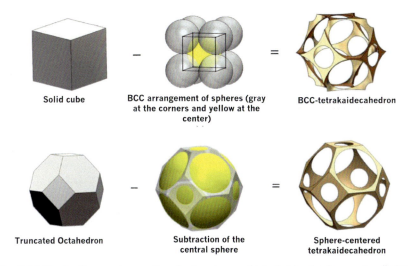

Fig. 2 Methods of geometry creation for two types of tetrakaidecahedron unit cells [17].

note, however, that despite these variations, the unit cells in metal foams exhibit a stochastic or randomly organized arrangement in three dimensions. Furthermore, the strut interconnections include a "blob", a feature absent in the theoretically obtained unit cell models. This nuanced aspect is discussed in more detail in a subsequent section.

The major focus of earlier investigations on stochastic metal foams was on studying the flow and thermal properties of these materials, with an emphasis on a comprehensive evaluation of permeability (K), inertial coefficient (C_f), effective thermal conductivity (k_{eff}), and interstitial heat transfer coefficient (h_{sf}).

The pressure drop per unit length (Eq. 1) for flow through a channel fully occupied by metal foam depends on the flow velocity and the metal foam properties, e permeability and inertial coefficient. These, in turn, depend on the porosity, and a relationship between the strut and pore diameters (which in turn depends on the porosity) [18].

$$-\frac{\partial p}{\partial x} = \frac{\mu \langle u \rangle}{K} + \frac{\rho C_f}{\sqrt{K}} \langle u \rangle^2 \qquad (1)$$

$$K = d_p^2 \left[0.00073(1-\varepsilon)^{-0.224} \left(\frac{d_f}{d_p}\right)^{-1.11} \right] \qquad (2a)$$

$$C_f = 0.00212\left[(1 - \varepsilon)^{-0.132}\left(\frac{d_f}{d_p}\right)^{-1.63}\right]$$

(2b)

where the strut-to-pore diameter ratio is given as,

$$\frac{d_f}{d_p} = 1.18\sqrt{\frac{1 - \varepsilon}{3\pi}}\frac{1}{(1 - e^{-25(1-\varepsilon)})}$$

(3)

To determine the strut diameter (d_f) for a given foam sample, high spatial resolution imaging with an accurately defined length scale is employed. The porosity (ε) of the sample is then ascertained by weighing the sample, measuring the total foam volume, and utilizing the known values of the densities of the metal and air. Leveraging this information, the pore diameter (d_p) can be determined using Eq. (3), while the values of K and C_f are easily calculated from Eq. (2).

Alternatively, one can ascertain the values of K and C_f by directly measuring the pressure drop across a finite length of metal foam at different flow velocities. A regression analysis of this data can then be used to extract the two coefficients, K and C_f, appearing in the quadratic relationship between pressure gradient and flow speed, as expressed in Darcy-Forchheimer equation (Eq. 1). In the context of buoyancy-driven convection, in which flow velocities are typically low and the pressure gradient in the direction of bulk flow is primarily governed by Darcy Law, the permeability of the foam sample plays a major role in identifying foam configurations that exhibit superior heat dissipation capabilities. Generally, foam samples with higher permeability tend to result in enhanced heat dissipation capabilities [1]. Conversely, in applications requiring high rates of heat removal, where flow is driven by an external fan and the flow velocities are generally high, both K and C_f become pertinent in determining the net pressure drop or pumping power required to achieve a specific level of heat dissipation.

The K and C_f values of the metal foam samples have been used to determine the net pressure drop of different metal foam volumes as well as in the computational studies based on volume-averaged (VAR) approach. The governing equations for mass and momentum conservation for steady-state conditions following the VAR approach are given as,

$$\frac{\partial(\rho v_j)}{\partial x_j} = 0$$

(4)

$$\frac{\partial\left(\rho\frac{v_i v_j}{\varepsilon^2}\right)}{\partial x_j} = -\frac{\partial p}{\partial x_i} + \frac{\partial}{\partial x_j}\left[\frac{\mu}{\varepsilon}\left(\frac{\partial v_i}{\partial x_j} + \frac{\partial v_j}{\partial x_i}\right)\right] - \left\{\frac{\mu\langle u\rangle}{K} + \frac{\rho C_f}{\sqrt{K}}\langle u\rangle^2\right\}$$

(5)

Now consider the energy transport Eqs. (6)–(8), where effective thermal conductivity, k_{eff}, and heat source generation term S_h, pertinent to metal foams, arise. Under local thermal non-equilibrium and absence of internal heat generation in metal foam volume, the energy transport equation for the solid and fluid phases in steady state are given as,

$$\frac{\partial(\rho v_i c_p T_f)}{\partial x_i} = \frac{\partial}{\partial x_i}\left[k_f\left(\frac{\partial T_f}{\partial x_i}\right)\right] + S_h$$

(6)

$$0 = \frac{\partial}{\partial x_i}\left(k_{eff}\frac{\partial T_s}{\partial x_i}\right) - S_h$$

(7)

where, the source term $S_h = h_{sf} a_{sf}(T_s - T_f)$. The interstitial heat transfer coefficient (h_{sf}) is given as,

$$h_{sf} = \frac{m Re_{d_f}^n Pr^{0.37} k_f}{d_f}$$

(8)

where, Re_{d_f} is the Reynolds number based on the strut diameter (d_f), Pr is Prandtl number, m $= 0.76$, n $= 0.4$ for $Re_{d_f} \in$, [3,40] m $= 0.52$, n $= 0.5$ for $Re_{d_f} \in [40,10^3]$, and m $= 0.26$, n $= 0.6$, for $Re_{d_f} \in [10^3, 2 \times 10^5]$ [1,19]. The surface area-to-volume ratio (a_{sf}) is given as $3\pi d_f G / d_p^2$, where G is the "shape function" given as $(1 - e^{-25(1-\varepsilon)})$ [18,20]. Note that for local thermal equilibrium cases, $S_h = 0$ as $T_s = T_f$. Hence, an accurate knowledge of k_{eff} and h_{sf} is imperative to simulate thermal transport in metal foams through the VAR approach.

Mahajan and co-researchers [19] proposed a theoretical model for effective thermal conductivity by introducing a geometry-based semi-empirical relationship. In their study, a typical unit cell representative of the stochastic metal foam (Fig. 3) was considered that incorporated a two-dimensional model with strut interconnections, including the blob (approximated as a rectangle) observed in actual metal foam samples. It was shown that the effective thermal conductivity is a function of the geometrical parameters, metal foam porosity, and intrinsic thermal conductivities of solid and fluid phases. Validation against experimental data

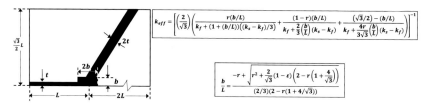

Fig. 3 Two-dimensional unit cell model considered in [19] and corresponding k_{eff} expression.

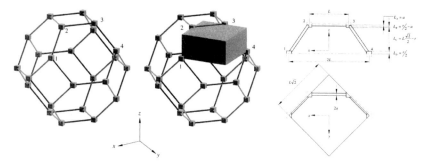

Fig. 4 The TKD model with cylindrical struts and cubic interconnections and a 2D unit cell [21].

collected for air and water-based experiments demonstrated the effectiveness of the presented k_{eff} relationship (Fig. 3), showing a good fit for $r = 0.09$, where r is a geometric parameter defined as the ratio of the thickness of the fiber to the width of the bump. In a follow-up study, Bhattacharya et al. [18] argued that a circular approximation of blob found at the strut interconnections is more suitable. A similar two-dimensional representative unit cell was modeled. The proposed semi-empirical relationship proved to be a good fit with the experimental data for $r = 0.19$, where "r" now was the ratio of the half thickness of the fiber to radius of the circular intersection. Another geometry-based k_{eff} calculation was presented in [21] (Fig. 4) and a revised correlation was prescribed by Dai et al. [22].

Regarding the interfacial heat transfer coefficient, h_{sf}, Calmidi and Mahajan [23], derived this parameter from their heat transfer measurements, assessing the heat dissipation capabilities of aluminum foams with varying pore densities (5–40 ppi) and porosities (0.9 to 0.97). These foams were subjected to air forced convection. The constant $m = 0.52$, in Eq. (8) was found to be appropriate. Their data are summarized in Fig. 5. The

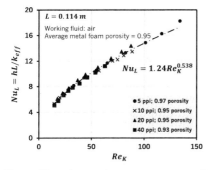

Fig. 5 Nusselt number versus permeability-based Reynolds number variation for aluminum foams with different pore-densities [23] (forced convection with air as working fluid).

convective heat transfer coefficient was presented in a normalized form by considering the total heat length in streamwise direction and the effective thermal conductivity of respective metal foams. The Nusselt number (hL/k_{eff}) was found to have a power-law dependence on the permeability-based Reynolds number (Re_K). Here we present a single heat transfer equation ($Nu_L = 1.24 Re_K^{0.538}$) based on permeability-based Reynolds number which adequately fits different pore densities, with average porosity of 0.95. In [23], h was based on the length-averaged wall temperature and the inlet fluid temperature. An alternative approach to presenting h would be to determine local values in the streamwise direction based on local wall temperature and local fluid temperature. The length-averaged heat transfer coefficient can then be determined by integrating the local values of h over the streamwise length.

High-porosity aluminum foams have also been experimentally characterized by Bhattacharya and Mahajan [24] for buoyancy-driven convection for a wide range of Rayleigh numbers. Experiments were carried out for two different heating orientations: one involving lateral heating (vertical position) of the base plate on which the metal foam was mounted, and the other with the heating plate was maintained horizontally. The obtained results, shown in Fig. 6 illustrate the variation of the Nusselt number with Rayleigh numbers for different pore densities with an average porosity of 0.94, under both heating orientations. For the horizontal heating orientation, the stream function (ψ) and the normalized fluid temperature (θ_f) are also shown. Collectively, the figure provides insights into how the Nu changes with Ra, offering a comparative analysis across diverse pore density metal foams and two commonly encountered heating orientations.

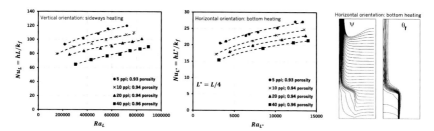

Fig. 6 Nusselt number versus Rayleigh number variation [24], along with the stream function and normalized fluid temperature variation in the metal foam volume and the enclosure [1] (buoyancy-driven convection with air as working fluid).

As evident from Fig. 6, high porosity metal foams can be very effective in buoyancy-driven convection scenarios which require no external pumping power to facilitate the convection. In contrast to forced convection, applications reliant on buoyancy-driven convection favor low pore density foams due to their higher permeability, which enables an unimpeded upward rising density-difference driven flow through the metal foam pores. Bhattacharya and Mahajan [25] also investigated a novel finned metal foam concept for electronics cooling applications for both forced [25] and buoyancy-driven convection [24] scenarios, where thin metal foam inserts were placed in between adjacent longitudinal fins of a conventional heat sink.

In addition to the investigations discussed above on both forced and buoyancy-induced convection in metal foams with air aw working fluid, numerous researchers have examined the transport phenomena in foams for another commonly encountered fluid: water. For instance, Hunt and Tien [26] conducted experimental studies on heat transfer through high porosity carbonaceous and metal foams under forced convection with water as the working fluid. Their investigation encompassed the characterization of carbon, nickel, and aluminum foams.

Fig. 7 presents their data plotted as length-averaged effective thermal conductivity-based Nusselt number vs permeability-based Reynolds number, with details of the metallic foam's properties provided in the inset Table. We are presenting the normalized heat transfer (with effective thermal conductivity) versus permeability-based Reynolds number, based on the data provided in [26]. It was found that the three distinct trendlines for vastly different solid-phase thermal conductivities collapsed onto one single curve ($Nu_L = 92.3 Re_K^{0.69}$). Compared to air-based convection studies conducted by Calmidi and Mahajan [23] presented earlier, the water-based

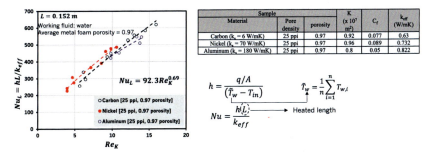

Fig. 7 Nusselt number versus permeability-based Reynolds number variation for carbon, nickel and aluminum foams [26] (forced convection with water as working fluid).

cooling resulted in significantly higher convective heat transfer coefficient. This outcome is an expected outcome, considering water's thermal conductivity is ~25 times higher than that of the air.

High porosity metal foams have also been tested in their compressed form for highly compact water-based convection heat exchangers by Boomsma and Poulikakos [27]. Fig. 8 shows a 4x compressed metal foam and the heat transfer characteristics of metal foams compressed by different factors.

Compressed metal foams were found to be very effective in heat dissipation for a given water flow rate, as well as in terms of the reduction in thermal resistance at a given pumping power to maintain a certain flow rate across the heat exchanger. High porosity metal foams have been widely studied in different applications; a comprehensive review on such studies can be found in [15].

Although metal foams have been proven to be effective enhanced heat transfer agents, their widespread application has been largely limited due to many factors, including limited pore density and porosities they are manufactured in, the types of materials these metal foams are available in, and the manufacturing process itself, which yields in unit cell of a particular type (based on minimum free surface energy principle). Moreover, adapting these metal foams for diverse applications, particularly in compact heat exchangers, presents challenges. Traditional machining methods, like cutting or milling, tend to deform the intricate struts and pores of the metal foam. Standardized form factors such as cubes, cuboids, or cylinders, further constrain their versatility. One other major concern of metal foams is the contact resistance when it is attached to the walls from where heat must be exchanged. Generally, a thermally conductive paste is applied between

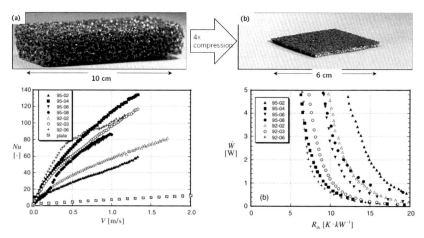

Fig. 8 Compressed metal foams' heat transfer and thermal hydraulic performance [27].

the metal foam sample and the heated wall to reduce thermal resistance. However, the thin struts have minimal contact with the heated wall, which renders the application of thermal paste ineffective. Metal foam manufacturers often offer brazed samples onto a substrate, but this approach proves costly, non-scalable, and adds further limitations on the form factor.

2.2 Looking ahead

Over the past two decades or so, researchers have been motivated to develop porous media concepts that retain the advantages offered by metal foams while addressing the associated challenges. Another motivation of latest research in this area is to improve metal foam functionality by incorporating mechanical strength and robustness aspects into the heat exchanger design framework. Advancements in manufacturing science, a multifold increase in computational speed and reduction in computational costs have enabled non-intrusive topology characterization and 3D volume reconstruction using advanced CAD modeling software. This transformative progress in porous media research has yielded diverse topologies and innovative heat exchanger concepts.

As introduced earlier, cellular materials can be broadly classified in two categories: stochastic and architectured. Further, we divide architectured foams into three sub-categories, each characterized by the nature of the solid-phase topology within a typical unit cell. These subdivisions are strut-based, sheet-based, and planar. It is important to note that even the

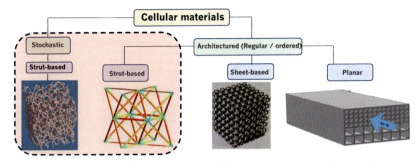

Fig. 9 Classification of cellular materials for thermal transport applications [17,28–30] and the focus of the present article (highlighted in red).

stochastic foams fall under the strut-based category, for which a typical strut of a TKD unit cell, as shown in Fig. 2, is of tricuspid shape [17]. A detailed presentation of the mechanical characteristics of different types of cellular materials, shown in Fig. 9, has been covered by Ashby [31] and Wadley [32] in 2006. For the sake of brevity and focused discussion, the subsequent text exclusively delves into the strut-based cellular materials, both stochastic and architectured cellular materials, as highlighted in red in Fig. 9, pore-scale flow and thermal transport in them and the engineering advancements associated with these specific materials.

3. Pore-scale flow and thermal transport in stochastic cellular materials

Earlier studies, as discussed in Section 2 regarding stochastic metal foams, have demonstrated the highly effective nature of metal foams in heat dissipation. A combination of a hybrid experimental-computational framework was employed to gain an understanding of the underlying thermal transport mechanisms. The computational aspect of these studies adopted a VAR approach, relying on empirical correlations for flow and thermal transport quantities such as, K, C_f, h_{sf}, k_{eff}. However, the computational analysis had limitations, particularly in its adoption of a two-dimensional (2D) unit cell-based approach for k_{eff} calculations. This approach, while insightful, fell short of authentically capturing the intricate unit cell topology inherent in stochastic foams and its nuanced effect on local flow and thermal transport. These 2D models, employing a combination of series and parallel conduction paths in solid and fluid phases, were constructed based on experimental data. However, this approach provided only a limited insight into local flow physics and thermal transport in stochastic metal foams. The reliance on experimentally obtained flow and thermal transport quantities posed challenges, as these measurements captured bulk phenomena. Consequently, gaining a fundamental understanding of micro-scale resolved fluid dynamics and achieving highly accurate determination of thermal quantities remained elusive.

Around 2005, a pivotal shift occurred with researchers delving into pore-scale fluid dynamics, using theoretically determined unit cell structures with appropriate computational boundary conditions. Krishnan et al. [33] carried out direct simulations on a Body-Centered Cubic (BCC) unit cell, as depicted in Fig. 2. They leveraged flow and thermal periodicity to mitigate computational cost associated with simulating a large number of unit cells representative of the stochastic metal foam volume. This innovative approach yielded detailed velocity and temperature fields at the unit cell level, significantly enhancing the understanding of fundamental thermal transport in metal foams. While, the simulation results for the BCC unit cell in [33] were highly accurate, it is noteworthy that the unit cell itself was idealized. Stochastic foams, on the other hand, are characterized by randomized arrangements of TKD shaped unit cells, featuring local defects and strut-interconnections of different geometrical shapes. This realization motivated further investigations in pursuit of more realistic unit cell models to facilitate comprehensive computational studies.

In a subsequent investigation conducted by the same research group, Bodla et al. [34] carried out direct computations on a realistic domain derived from micro-tomography of actual metal foams, followed by image processing and additional steps required for mesh generation. Employing a spatial resolution of $20\,\mu m$, the study focused on the largest possible test article which could be scanned ($10\,mm \times 12.7\,mm \times 38.1\,mm$) using a commercial X-ray scanner. To accomplish this, a larger volume of metal foam was precisely cut using Electrical Discharge Machining to generate the above foam volume. This volume featured a sufficient number of unit cells randomly arranged in three dimensions. This arrangement allowed for the simulation of a computational domain with velocity inlet and pressure outlet type boundary conditions, complemented by symmetry boundary conditions on the remaining faces.

For simulations pertaining to the effective thermal conductivity (k_{eff}), a larger computational domain with a higher number of unit cells was favored, driven by the inherent random organization of pores in metal foams. This approach aimed to ensure that the reported k_{eff} values remained representative across various pore arrangements, mitigating the influence of specific configurations. The computed effective thermal conductivity obtained by this approach was then compared with established models, such as those proposed by Calmidi and Mahajan [19], Bhattacharya et al. [18], Boomsma and Poulikakos [21], Lemlich [35]. The results of this comparison, shown in Fig. 10, reveal a general convergence of prior models to similar values, particularly for larger porosity values, excluding [21]. This convergence is attributed to the diminishing influence of the solid-phase or strut-modeling approach at higher porosities. Bodla et al.'s simulations [34] specifically carried out for porosities lower than 0.92, exhibited notable variance within the narrow range of investigated porosity. Despite this variance, the findings generally align with the models presented in [18,19,35]. Some other notable studies on k_{eff} calculations were also conducted in [36,37].

Ranut et al. [28] conducted pore-scale computations on a fluid domain generated by micro-CT scanning and subsequent processing. These computations focused on incompressible laminar flow within 10–30 ppi metal foams.

The resulting computationally determined permeability values were compared against a larger number of prior experimental and numerical results. It is noteworthy that the computed value displayed in Fig. 11 (shown by symbol x) is an average of the three directional permeability values. This aligns with the expectation of a unique permeability value for a

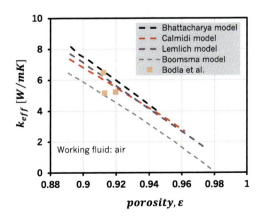

Fig. 10 Comparison of effective thermal conductivity obtained on a metal foam model generated using μ-CT data with the known correlations.

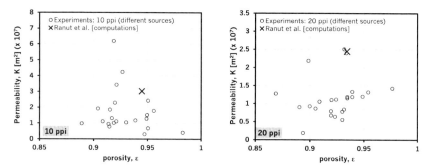

Fig. 11 Comparison of pore-scale modeling-based permeability determination with different experimental sources as reported in [28].

foam of given pore density and porosity. However, this ideal scenario was not consistently observed when comparing permeability across different experimental studies. Notably, a large variation in permeability values was observed at a porosity level of ~0.92 for both 10 and 20 ppi foams. A key contributor to the observed variability in the experimentally determined permeability values is the selection of the maximum flow velocity used in the regression analysis between pressure gradient in the flow direction and average flow velocity. Boomsma and Poulikakos [38] were the first ones to highlight this concern in the processing of flow data, emphasizing that permeability and inertial coefficient are inherent properties of metal foams and should ideally be independent of the flow conditions or the chosen maximum flow velocity for data reduction. Our in-house investigations on architectured materials confirm the findings of Boomsma and Poulikakos [38], underscoring the need for careful consideration and standardization in the methodology employed for permeability determination to ensure consistency across experimental studies.

In the last decade, pore-scale computational studies have primarily concentrated on resolving local flow features at the pore and strut levels, with a predominant focus on simulations conducted under laminar flow conditions. The governing equations for mass, momentum, and energy conservation for steady laminar flow through metal foams for direct simulations are as follows:

$$\frac{\partial}{\partial x_i}(\rho u_i) = 0 \tag{9a}$$

$$\frac{\partial}{\partial x_j}(\rho u_j u_i) = -\frac{\partial p}{\partial x_i} + \frac{\partial}{\partial x_j}\left(\mu \frac{\partial u_i}{\partial x_j}\right) \tag{9b}$$

$$\frac{\partial}{\partial x_i}(\rho c_p T u_i) = \frac{\partial}{\partial x_i}\left(k_f \frac{\partial T}{\partial x_i}\right) \qquad (9c)$$

It is noteworthy that this approach circumvents the reliance on semi-empirical correlations for flow and heat transfer, making it applicable to any topology for laminar and steady-state conditions. Pore-scale computational efforts can be broadly classified into two methods based on the approach to geometry creation: (1) 3D flow and solid domain obtained by post processing 2D images obtained from μ-CT scanning and subsequent reconstruction, and (2) theoretically constructed representative unit cell and its reticulation in three dimensions to obtain the 3D flow and solid domain, which can effectively represent the actual volume of metal foams. In this sub-section, our focus is on stochastic geometries, deploying method 1, with the discussion on ordered geometries resulting from the 2nd method deferred to later sections.

Regarding the 1st method, a sample process for 3D volume generation and meshing is shown in Fig. 12.

This method adeptly captures the intricate details of stochastic foams, providing a precise representation of unit cells, their randomized placements, strut interconnections, the strut junction shapes, and any defects. Although it is imperative to capture these details for an accurate simulation of a 3D model that aligns with experimental observations, the presence of rough struts with sporadic defects makes the grid generation for finite-volume-based solvers challenging. In addressing these challenges, post-processing becomes essential to handle sharp solid volumes, necessitating smoothing techniques, particularly in cases where high curvatures are detected at strut junctions. Although these simplifications are essential to conducting detailed computations, they may inadvertently introduce minor deviations from experimental findings. Nevertheless, even with these adjustments, most of the local flow and heat transfer details are captured with high accuracy in the 3D models generated through this method, serving as the foundation for direct numerical simulations.

Diani et al. [40] performed pore-scale simulations to study the local fluid dynamics of flow through high-porosity copper foams. Utilizing a $20\,\mu m$ CT scanning, they acquired 2D images employed in the generation of 3D volume for grid generation. A considerably large physical domain was simulated for a comprehensive understanding of the flow's evolution as it moves through the pores. The number of elements for 5 to 40 ppi foam

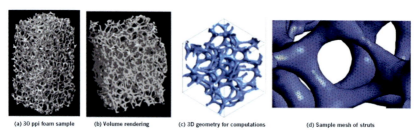

Fig. 12 Process of 3D volume generation from μ-CT scan images [39].

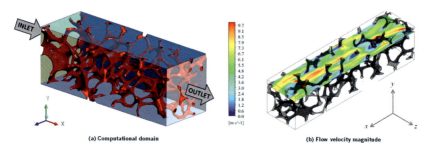

Fig. 13 (A) Computational modeling domain, (B) velocity magnitude contour [40].

geometries ranged d between 3 and 27 million. Fig. 13 shows the modeling domain and velocity magnitude contour for 10 ppi foam used in [40].

It can be observed that flow velocities are high in regions with high local permeability, while low velocity zones are typically observed in the wake region of struts.

More recently, μ-CT scanning, coupled with advanced software like ImageJ and Mimics Research, has been employed for fine tune probing of the strut morphology to obtain an almost exact skeletal structure compared to the physical foam sample. For instance, Yu et al. [41] utilized a μ-CT scanning resolution of up to 0.4 μm to capture micropores within the struts, see Fig. 14. Their study employed the realizable k-ε model to investigate the impact of these micropores on both local and overall flow and heat transfer in turbulent air flow. The results indicated that the skeletal structure with micropores experienced higher pressure drop (~11%) compared to the case of a solid skeleton (absence of micropores).

In a follow-up study [42], the authors investigated the effect of modeling the micropores on net heat transfer. The findings revealed that the model incorporating micropores demonstrated a marginally higher

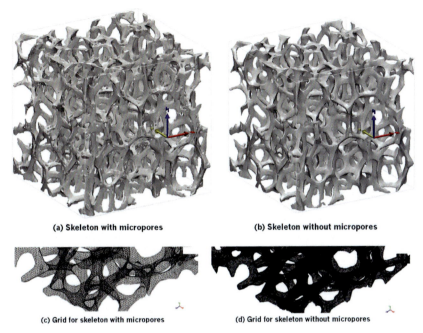

Fig. 14 Ultrahigh resolution μ-CT scan-based 3D volume reconstruction to resolve micropores [41].

convective heat transfer compared to the model without micropores. These recent studies [41–43], where metal foam struts were resolved at very fine length scales, underscores that previous analyses relying on oversimplified geometries may not have adequately captured the detailed fluid dynamics at the strut and pore levels. Consequently, predictions from the older models, particularly for permeability, may not be very accurate due to the aforementioned limitations. In some other recent investigations [44,45] on turbulent flows through stochastic foams, the realizable k-ε treatment for turbulence modeling was incorporated. Although accuracy in modeling the fluid domain was achieved, the Reynolds-Averaged-Navier-Stokes-based turbulence modeling approach was noted for its limitations in providing accurate flow and heat transfer quantities. Further, research in this domain may involve direct numerical simulations on stochastic models generated via high spatial resolution μ-CT scanner.

The second category of investigations concerning stochastic foams involves generating fluid and solid domains through idealization and/or smoothing of the struts while still preserving their random organization.

This approach provides an economical means to understand pore-scale flow dynamics without the need for detailed μ-CT scanning and its associated post processing. Furthermore, while μ-CT scanning reveals intricate geometrical features, it is constrained by the maximum volume of the sample size that can be scanned, limiting the examination of larger foam volumes. To this end, leveraging an understanding of typical unit cell topology encountered in the foaming process-enabled stochastic foams, and knowledge of the relative arrangement of such topologies in 3D volume enables the generation of large metal foam volumes with idealized or approximated strut shapes and their junctions. Fig. 15 illustrates an example of the approximation of 3D geometry obtained by μ-CT scan images processing [43]. Such approximated geometries also allow mathematical transformations, which could result in different geometrical configurations while preserving the basic nature of the unit cell and their interconnections in three dimensions. Various other methodologies have been recently adopted to generate stochastic representative metal foam volumes that exhibit properties akin to actual metal foams.

Regardless of the method employed to approximate intricate geometrical features, the fundamental principle guiding geometry creation is the minimization of free surface energy. This principle is essential to preserve the basic pore and strut geometry of metal foams produced via the foaming process. In 1887, Thomson [46] proposed the TKD structure to satisfy this principle (also known as Lord Kelvin unit cell). A century later, Wearie and

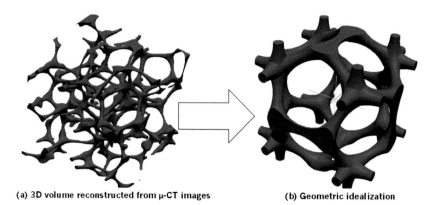

Fig. 15 (A) 3D volume reconstructed from μ-CT imaging, (B) an idealized geometry with scaling [43].

Phelan [47] proposed another unit cell which had 0.3% lower surface energy compared to the TKD model.

Both models, as shown in Fig. 16, are based on the formulation presented in [48]. Kopanidis et al. [48] adopted this method of strut generation, and observed that it yields a triangular cross-sectional strut with an inner concavity. A similar geometry creation approach was adopted in [49] where Lattice Boltzmann simulations were conducted to resolve the local flow and heat transfer. In their work, Kopanidis et al. [48] applied a finite-volume approach on the geometry shown in Fig. 16C, and considered two types of thermal boundary conditions for the struts: (1) where bottom plate and the struts were subjected to constant temperature, and (2) conjugate heat transfer where only the bottom plate was maintained at a fixed wall temperature. The second scenario, shown in Fig. 17, incorporated conjugate heat transfer conditions, which represent the more realistic thermal boundary conditions for metal foams. The highly detailed nature of the geometry and well-defined struts facilitated the resolution of pore-scale flow- and temperature-fields, enabling the precise determination of heat diffusion within the struts.

This is evident in the thermal gradients observed in the struts, with temperature generally decreasing with increasing distance from the heated substrate. It is also expected that the strut-level heat transfer coefficients will differ at the interface with the bottom wall compared to locations farther away. This effect, combined with the diminishing capacity for heat transfer in the mid-channel zone (attributed to a decrease in the difference between local strut and fluid temperature), results in notable variations in local convection heat transfer levels. This is an important consideration which should be accounted for in the VAR approach, where a fixed interstitial

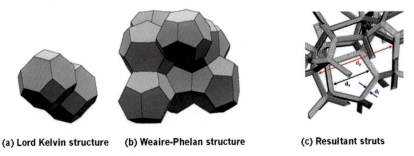

(a) Lord Kelvin structure   (b) Weaire-Phelan structure   (c) Resultant struts

Fig. 16 Pore-scale geometry creation methods via Lord Kelvin and Weaire-Phelan approaches [48].

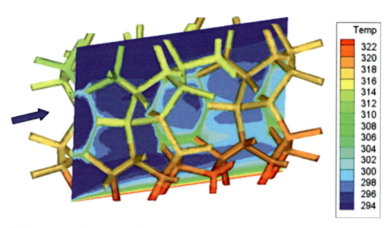

Fig. 17 Conjugate heat transfer characteristics of struts attached to a base plate subjected to constant wall temperature [48].

heat transfer coefficient is uniformly applied to the entire solid phase irrespective of its proximity to the base plate or distance from the channel inlet (Fig. 18).

The geometry creation method described in [48] was later adopted by Moon et al. [50] who delved into even finer details of actual metal foam geometry, uncovering the presence of hollow struts at sporadic intervals. Such distinctive geometrical characteristics of actual metal foam can only be realized through an engineered approach, involving the intentional design of pore and strut geometry arranged stochastically in three-dimensions. The authors [50] carried out a parametric study on the hollowness factor defined as the ratio of empty area to the total cross-sectional area of a strut (A'/A). Their findings revealed that the net heat transfer decreases with an increase in hollowness factor. They emphasized the importance of considering this aspect in computational studies aiming to validate predictions against experimental findings.

An alternative approach to generating representative geometries of stochastic metal foams, involving the generation of a sphere-void-phase, is demonstrated in [51,52]. Although the resulting geometry may not be entirely stochastic in nature, as a specific strut network repeats itself in three dimensions, the stochastic characteristics are retained within local groups.

Recently, a novel approach based on Laguerre-Voronoi tessellation (LVT) was employed for generating representative metal foam volumes [53]. This process involves the creation of a bounding volume and random

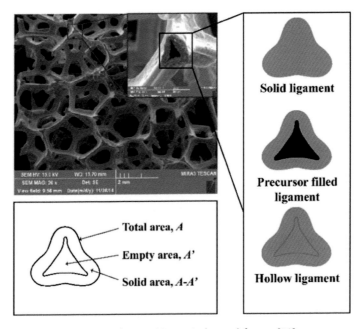

Fig. 18 Strut hollowness as observed in typical metal foams [50].

placement of seed points using the LVT algorithm, followed by generating spheres centered at these randomly placed seed points (Fig. 19). Employing a parallel Lattice-Boltzman Method (LBM), the study delved into the conjugate heat transfer characteristics of metal foams with porosities varying between 0.8 and 0.9 and pore density from 10–30 ppi. Another recent study on LVT was carried out in [54]. More studies are expected in this domain, as it allows the engineering of stochastic metal foams by providing control over the unit cell creation at user-specified cell centers.

4. Pore-scale flow and thermal transport in architectured cellular materials: strut-based

Over the last two decades, there has been a notable surge in research on characterizing pore-scale flow and thermal transport within architectured cellular materials. Advances in manufacturing methods have provided design freedom leading to the creation of innovative unit cell topologies with multi-functional attributes, not limited to just superior thermal hydraulic performance. The common manufacturing methods

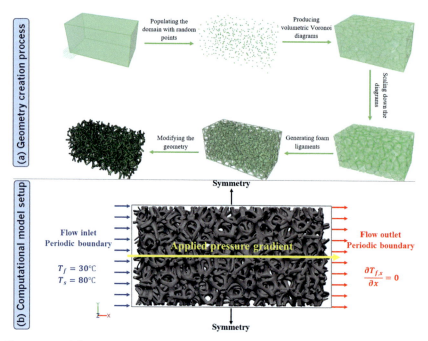

Fig. 19 Metal foam representative geometry creation by Laguerre-Voronoi tessellation and the boundary conditions for LBM computations [53].

employed for porous media (apart from the foaming process discussed earlier) include metal-wire approach, investment casting, sheet metal bending, and more recently metal additive manufacturing (AM). In particular, metal AM has made significant strides in the last decade, which have considerably expanded the design possibilities of constituent unit cell topologies in metal foams. This advancement has also enhanced the feasibility of manufacturing alloys capable of withstanding high temperatures. The above manufacturing methods can yield cellular materials of different types such as strut-based, sheet-based, and planar, see Fig. 9. For a comprehensive review on these manufacturing methods pertaining to regular or architectured lattices, the readers are referred to Kaur rand Singh [55]. In this section, we provide a thorough overview of some innovative unit cell topologies realized through the aforementioned manufacturing methods. Additionally, we present corresponding computational investigations aimed at understanding the pore-level flow and thermal transport.

4.1 Forced convection with air as working fluid

In this sub-section, we present forced convection studies on architectured materials using different working fluids such as air, water, supercritical carbon dioxide, particles (serving as convection transport agent and referred as fluid), and hydrocarbons. Some of the earlier investigations employed the metal-wire approach (Fig. 20), utilizing copper wires with a thermal conductivity of 385 W/mK and diameters ranging from 0.6 to ~1 mm [56]. Although this technique is effective in heat dissipation due to its highly conductive solid phase and good contact with the base plate, it shares a limitation with the foaming process-based metal foams. Specifically, this manufacturing method cannot be extended to generate more complex strut-based topologies. Additionally, when aiming for smaller length scale struts, desirable for the development of compact and efficient heat exchangers, the metal-wire approach encounters challenges that must be addressed.

The metal-wire woven technique was adopted to obtain Kagome-type cellular structure in a sandwich configuration where the solid-phase was aluminum and the cellular structure was brazed onto the base plates [57]. A typical sample and its representative unit cell are shown in Fig. 21. The

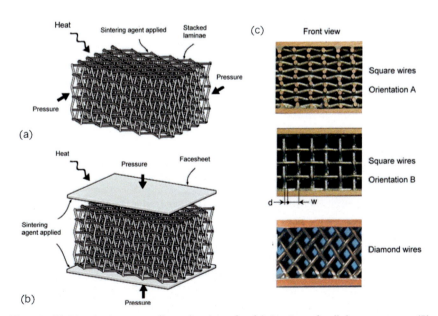

Fig. 20 (A) Metal-wire or textile technology for fabrication of cellular structures, (B) bonding the cellular structure with base plates which will be subjected to heat load, (C) two orientations of square wires and one for diamond wire [56].

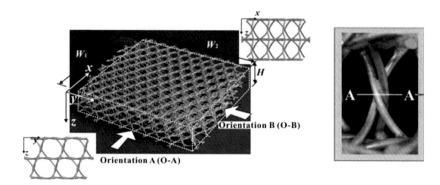

Fig. 21 Sandwich configuration of Kagome shape-based cellular structure through wire-weaving technology, and a representation unit cell [57].

thermal conductivity of solid phase was 222 W/mK and the strut diameter was ~1 mm. Experiments were conducted using air for both laminar and turbulent flow regimes. The convective heat transfer levels achieved were comparable to those observed in investment casting-based test articles that produced single-layer Kagome-shape based sandwich structures. The Nusselt number, based on channel hydraulic diameter, varied between 10^2-10^3 for Reynolds numbers between 10^3 and 10^4 [57]. The wire-weaving technique, resulting in typical Kagome-shaped structures, has been further explored by several researchers, primarily in sandwich panel configuration with a single layer of Kagome unit cells (as shown in Fig. 22). Kagome unit cells comprise of six struts joined together at the center of the unit cell volume, forming a body-centered type. The thermal transport mechanism involves effective conjugate heat transfer, as the struts transport heat from their connections with the base plates (subjected to heat load) to the junction, which offers stagnation or impingement-based effective convective action.

A detailed investigation on the flow-field of an inline arrangement of Kagome structures [59] reveal enhanced mixing resulting from a counter rotating vortex pair at a plane orthogonal to the bulk flow, flow stagnation at the strut portion facing the incoming flow, low velocity zones in the wake regions of the struts, and enhanced turbulence at the interface of struts and base plates. These distinctive flow characteristics, coupled with the aforementioned thermal transport mechanisms, provide substantial support to the observation that the advantages of Kagome shape-based configurations are more pronounced in sandwich configurations.

Thermal transport in engineered cellular materials

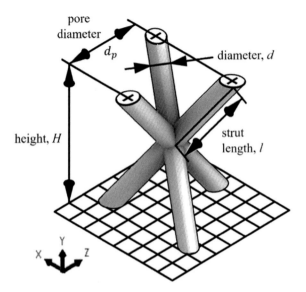

Fig. 22 A typical Kagome-shaped unit cell [58].

As demonstrated in a recent investigation [60], this configuration is particularly well-suited for cooling the trailing edge of gas turbine blades an application characterized by high heat loads, thermal stress, and constrained internal volume for implementing the cooling concepts. It stands out as a compelling alternative to the conventional pin-fin cooling systems [60].

Kagome-shaped unit cell can be modified by truncating the geometry from the mid-plane (parallel to bulk flow) intersecting the strut junctions and modifying the number of struts, as illustrated in Fig. 22. Two variations are presented in Fig. 23, adopting tetrahedral and pyramidal configurations. A comparative analysis of their performance, alongside Kagome and representative metal foam unit cell topologies [61], revealed noteworthy insights. Metal foam incurred the highest pressure drop, followed by tetrahedral and pyramidal structures, while Kagome demonstrated the lowest pressure drop amongst the investigated structures. In terms of heat transfer, the tetrahedral structure exhibited highest heat transfer followed by metal foam, pyramidal and Kagome. As commonly observed in enhanced heat transfer studies, the gain in heat transfer is inevitably accompanied by a simultaneous pressure drop. Therefore, it becomes imperative to evaluate these concepts based on their thermal hydraulic performance [62]. This aspect of design evaluation is particularly important for metal foams and architectured materials as they impose significant pressure drop due to their large wetted surface area interacting with the working fluid. Consequently,

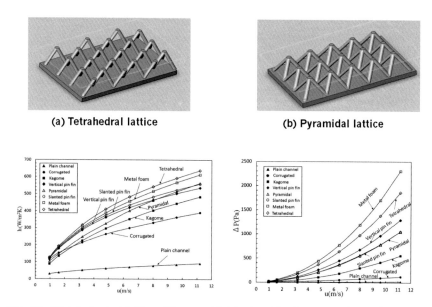

Fig. 23 (A) tetrahedral lattice, (B) pyramidal lattice. The heat transfer and pressure drop characteristics of different porous media concepts [61].

a comprehensive assessment considering both heat transfer efficiency and hydraulic performance is crucial in optimizing the design of such structures.

The superior heat transfer capabilities of tetrahedral topology arise primarily from the local enhancement in turbulent kinetic energy at the locations where struts intersect the endwall and face the incoming flow. Each individual strut can be viewed as a slanted cylindrical pin in a crossflow configuration. Analogous to the local heat transfer characteristics observed in pin configurations, the high heat transfer zones in tetrahedral structures correspond to the flow stagnation region, while comparatively lower heat transfer is observed in the strut wake regions [63]. In a follow-up study, the authors in [63], conducted experiments to determine endwall heat transfer using transient liquid crystal thermography (LCT), and static pressure measurements [64]. They also carried out computations on tetrahedral configuration to correlate the two horseshoe vortices generated at the respective endwalls and their corresponding nearby vortical structures, with the local endwall heat transfer. Demonstrated in these and various other studies [65–68], the application of LCT technique is highly effective in yielding high spatial resolution convective heat transfer coefficient maps at the endwalls. This method is particularly valuable in porous media research [64,69], since the junction between the porous media and the endwall is most critical for heat transfer. For an in-depth discussion of this measurement technique and the diverse applications of LCT, readers are referred to [70].

Following the experimental and computational investigations on Kagome, tetrahedral and pyramidal structures, exploration extended to the X-type lattice structure. This lattice arrangement can simply be obtained by folding a metallic sheet with preexisting cuts, offering a scalable fabrication process ideal for large-scale heat exchangers. In comparison with other similar structures, a single layer sandwich configuration of X-lattice has been studied, as depicted in Fig. 24, which shows the manufacturing process for pyramidal and X-type lattice, as well as the local Nusselt number characteristics of the X-type lattice, on the rear and front faces, for air as the working fluid. The X-type unit cell allows flow stagnation at the intersection, akin to the Kagome lattice. However, it distinguishes itself by providing enhanced heat transfer at the rear end of the struts due to a combination of spiraling flow and enhanced mixing [71]. For a given solid phase porosity, the X-type lattice yields significantly higher heat dissipation at a fixed pumping power in comparison to the Kagome structure [72]. Yan et al. [73] conducted an experimental and computational study on X-type lattice made from Stainless Steel 304 ($k = 16$ W/mK), where the lightweight structure was attached to thin base plates via thermal grease, and the base plate was subjected to constant heat flux via patch heaters. The superiority of the X-type lattice over the tetrahedral configuration was evident, with a ~75% higher Nusselt number at a given pumping power, partly attributed to the 66% larger surface area of X-type lattice. Recent applications of X-type lattices have extended beyond heat exchanger to disk braking [74] and modifications with roughness elements for even higher thermal hydraulic performance [75] (Fig. 25).

The architectured materials discussed so far can be manufactured via metal-wire, metal textile, sheet metal folding/bending, investment casting methods, which represent matured technologies that result in cost-effective and scalable heat exchanger solutions. While these methods allow for minor variations in configurations, for example metal sheet punching and then folding [76], control over the unit cells of architectured materials manufactured remains somewhat limited, similar to the foaming process. Future research on the design and development of next generation heat exchangers for industrial decarbonization through electrification of process heat [76], waste heat recovery [77], high temperature heat exchangers for gas turbines [78,79], concentrating solar power-based electricity

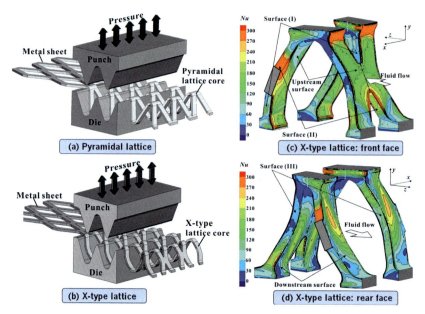

Fig. 24 Sheet metal folding fabrication process to obtain (A) pyramidal lattice, (B) X-type lattice [71].

production [80,81], hypersonic vehicles [82], requires multifunctional heat exchangers, which not only provide superior thermal-hydraulic performance, but also exhibit high mechanical strength, lightweight and cost-effectiveness.

The thermal management challenge is to strike a balance among the above-mentioned multifunctional attributes desired from the modern heat exchangers. In this, additively manufactured architectured cellular materials are poised to play a crucial role. With fast-paced advancements in metal AM processes and materials research for AM, the design flexibility offered through this manufacturing approach empowers heat transfer engineers to meet the diverse requirements of above-mentioned applications. With this perspective, we present in the following sections a comprehensive overview of research focused on AM-based cellular materials for enhanced heat transfer applications.

4.1.1 Strut-based additively manufactured lattice structures
4.1.1.1 BCC, FCC lattices and their variations
We first present the BCC structure, in the context of the Kagome structure discussed earlier. Recall the geometric relationship between the wire-woven

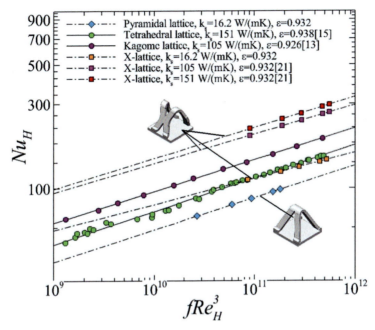

Fig. 25 Comparison of X-type, pyramidal, tetrahedral, and Kagome in terms of heat transfer versus normalized pumping power [72].

and Kagome structures, followed by the cylindrical-strut-based Kagome structure. In 2009, Wong et al. [83] were one of the first ones to use Selective Laser Melting (SLM) AM technology to fabricate complicated lattice structure. They conducted a comparative analysis of its flow and thermal performance, contrasting the results with those for conventional pin fins (also fabricated via SLM). More recently, Liang et al. [84] employed the transient LCT technique to compare the local heat transfer characteristics of a staggered arrangement of the BCC unit cell with Kagome and pin-fin, see Fig. 26, which elucidates the geometrical differences between the BCC and Kagome unit cell. Their findings revealed that the BCC lattice exhibited ~40% higher heat transfer compared to pin-fin and ~12% higher heat transfer compared to Kagome. Furthermore, the BCC lattice had the highest thermal hydraulic performance, followed by Kagome and pin-fins [84].

It is important to note that the transient liquid crystal techniques may not precisely isolate the true contribution of conjugate heat transfer. Instead, the resulting heat transfer data captures aspects of flow behavior, flow mixing, and near-wall turbulent kinetic energy. To comprehensively

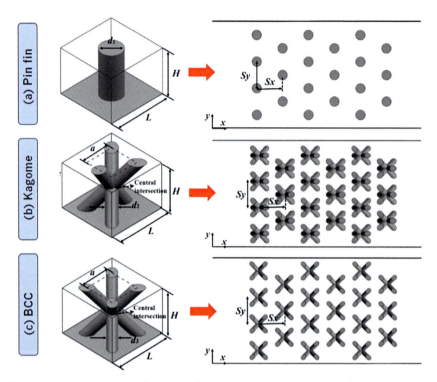

Fig. 26 A comparative study on pin fin, Kagome, and BCC lattices for their endwall local heat transfer characteristics [84].

understand the overall heat transfer properties of metallic lattice structures produced through AM, steady-state heat transfer experiments are crucial. These experiments involve heating the base plate is heated, allowing convective agent to carry away heat based on the structure's properties and prevailing flow conditions.

The enhanced heat transfer mechanism in metallic architectured cellular materials is a result of the synergistic effects of enhanced effective thermal conductivity, interstitial heat transfer coefficient, and endwall heat transfer coefficient. Subsequent to their initial study, the authors in [84] conducted steady-state heat transfer experiments on the three topologies shown in Fig. 26, where all the samples were additively manufactured by SLM [85]. In their investigation, they also included X-type lattices, which their data revealed had the highest heat transfer, surpassing the BCC lattice. However, the X-type lattice incurred a significantly higher pressure drop compared to the BCC lattice. In a related study, the same research group evaluated the thermal performance of an additively manufactured inline arrangement of two configurations based on Kagome unit cells in [86]. Although the heat transfer enhancement levels were 4–5 times higher than a comparable smooth channel (in absence of lattices), the concomitant pressure drop was also very high, rendering the concepts not very effective in the thermal-hydraulic performance spectrum where existing enhanced heat transfer concepts could be assessed and compared [86].

Pore-scale direct simulations were carried out in [87] for the BCC topology with porosity values between 0.7 and 0.99. The computational predictions revealed local velocity variation within the BCC unit cell, where pore-scale Reynolds number was kept below 150 [87]. As expected, structures with low porosity had significantly higher local flow velocities due to increased flow blockage compared to those with higher porosity. The effective thermal conductivity was also determined through this approach, and conclusions, similar to those in Section 2, were drawn, where it was shown that pore-scale simulations generally agree well with the prescribed semi-empirical correlations in high porosity regime (>0.9), but they deviate for lower porosities.

The BCC lattice was modified by adding additional struts and evaluated for its thermal performance in [88]. The study showed that the BCCZ configuration (8 +2 struts) resulted in the highest heat transfer for AlSi10Mg as the material for solid phase. The observed gain in heat transfer for the BCCZ configuration was more pronounced at lower porosities. In contrast, for higher porosities, various topologies, including BCC, FCC and their modifications demonstrated comparable heat transfer. It is worth noting that FCC lattices incurred higher pressure drop compared to BCC; the FCC struts directly obstructs the incoming flow through X-shaped orthogonal struts, whereas BCC struts are inclined, meeting at the center of the cubic unit cell instead of the face. The endwall heat transfer characteristics of FCC lattice was reported by Liang et al. [89], FCCZ heat transfer characteristics are reported in [90], and two-layer FCC structure placed in a duct was investigated in [91]. All of these studies were conducted for channel flows, where the convective agent flows through the entire lattice volume.

Porous media concepts also allow for other more efficient methods to route coolant through them, such as jet impingement at zero jet-to-target spacing. Singh et al. [20] investigated array jet impingement heat transfer performance for thin and high porosity aluminum foams manufactured via foaming process. This concept was later extended to single jet impingement onto a thick BCC lattice by [92].

4.1.1.2 Kelvin, Octet, face-diagonal cube, cube unit cell-based lattices

The BCC and BCCZ lattice structures discussed above exhibit high heat dissipation capabilities by channeling heat away from the base plates towards the junction at the center of the cube unit cell circumscribing the lattice, from where the thermal energy is transported via convection by the moving working fluid. Recent investigations in enhanced heat transfer under forced convection have delved into more complex unit cell topologies, such as the Kelvin cell with cylindrical struts (Fig. 27), Octet, Face-diagonal, and cube lattices. The Kelvin unit cells with cylindrical struts is an approximation of the body-centered or the sphere-centered TKD unit cell as shown in Fig. 2. AM enables the flexibility to choose strut topologies, with recent explorations on elliptical struts to lower the pressure drop and enhance thermal hydraulic performance [93]. The elliptical struts exhibited ~30% lower pumping power relative to the cylindrical struts [93]. Kelvin or TKD unit cells also serve as a baseline case for enhanced heat transfer studies focused on the development of novel unit cell topologies, as they are representative of the metal foams produced via foaming process Demonstrating the superior thermal-hydraulic performance over baseline topologies emphasizes the advantages of this manufacturing method.

To this end, the latest research on AM-enabled porous media is focused on complicated unit cells, such as those shown in Fig. 28, where one can observe the design freedom in unit cell topology, the number of sub-unit cells in a fixed volume, and the porosity of the unit cell itself, controlled by manipulating the strut diameter [94]. Another motivation behind the exploration of novel unit cell topologies is to develop multifunctional cellular materials, which not only have superior thermal transport

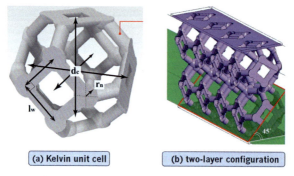

Fig. 27 Kelvin unit cell and a two-layer metal foam model generated by reticulating the unit cells [93].

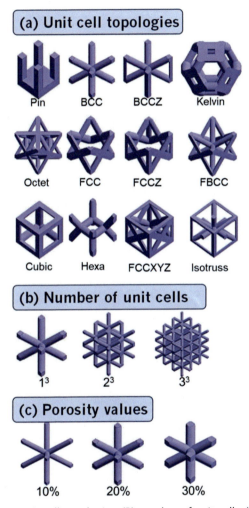

Fig. 28 (A) different unit cell topologies, (B) number of unit cells, (C) porosity values by changing the strut diameter [94].

properties, but also superior mechanical strength. Fig. 29 presents a comparison of Young's modulus, peak stress, and toughness of different strut- and sheet-based unit cell topologies [29]. Here, we will specifically discuss a comparison between the top two strut-based topologies, Octet and TKD (or Kelvin) unit cells. For high porosity (90%) samples, the Octet structure had significantly better mechanical strength compared to the TKD unit cell. Detailed investigations into the mechanical properties of Kelvin (TKD) and Octet unit cells can be found in [95,96].

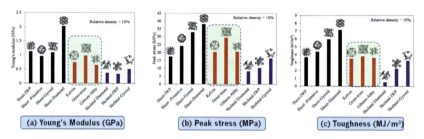

Fig. 29 Mechanical strength of different unit cell topologies, the strut-based topologies are highlighted in green [29].

A comprehensive experimental and computational investigation on Octet lattice was conducted by Chaudhari et al. [97] and Ekade et al. [98], respectively. Three different porosities ranging from 0.7 to 0.9 were additively manufactured in AlSi10Mg, and experiments were undertaken to determine the effective thermal conductivity, permeability, inertial coefficient and overall heat transfer coefficient of single-layer thick Octet lattice frame material [97].

The Octet lattice has been reported to have a lower pressure drop compared to stochastic foams. However, akin to the pore-scale computations carried out earlier by Krishnan et al. [99], the experimentally determined effective thermal conductivity in [97] did not agree well with the existing correlations widely used for metal foams. This persistent issue of the inapplicability of the existing semi-empirical correlations, particularly for stochastic metal foams for lattice porosities below 0.9, has recently been addressed by Kaur et al. [100]. In their work, a generalized correlation for effective thermal conductivity (Eq. 10) was proposed for both architectured cellular materials and stochastic foams with porosities above 0.8.

$$\frac{k_{eff}}{k_f} = 0.7(1-\varepsilon)^{1.22}(k_s/k_f) + 1 \tag{10}$$

where k_s, k_f are solid and fluid phase thermal conductivities respectively and ε is the lattice or stochastic foam porosity (>0.8).

Kaur and Singh [17] carried out pore-scale computations to understand flow and heat transfer in a unit cell configuration for Octet, Face-diagonal cube, TKD and Cube. The unit cell's edge length was 2.54 mm, allowing 1 in. (25.4 mm) to accommodate 10 unit cells. The porosity was kept fixed at 0.986 for all four topologies, which resulted in minimum strut diameter of 84 µm for the Octet topology. In Fig. 30, the interstitial heat transfer coefficient at the strut walls reveals that Octet offers multiple flow stagnation regions compared to the other three topologies, and its thin struts minimize the low heat transfer zones in the wake region. Although Octet and TKD had similar heat transfer areas, the volumetric Nusselt number for Octet was nearly 100% higher than that of TKD. This study underscored the potential of Octet topology through unit cell simulations, with the understanding that its true benefits can be realized through the three-dimensional reticulation of such unit cells and AM to form a porous media volume.

Kaur and Singh [17] conducted experimental investigation on the four topologies shown in Fig. 30. They additively manufactured a 5 × 1 × 5 (L x H x W) unit cell arrangement in a lattice frame configuration specifically, for gas turbine trailing edge cooling [79]. The samples were manufactured via Binder jetting AM technique, with Stainless Steel 420 (with 40% bronze infiltration). The AM technique chosen in [79] imposed several design constraints on the minimum strut diameter that could be accurately printed while maintaining the part's robustness. This constraint was notably stringent for the most complex topology (Octet), among the four shown in Fig. 30. Due to the large overhang of the struts in the Octet unit cell, the AM process permitted a 10 mm unit cell edge length and a resultant as-printed porosity of ~0.87 for each of the topologies. Note that 3D printing of lower porosity samples poses fewer challenges compared to high porosity

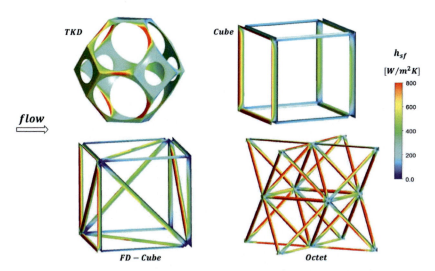

Fig. 30 Interstitial heat transfer coefficient of different unit cell topologies [17].

ones. However, beyond a certain lower limit of porosity, additional challenges related to material removal and post-processing emerge. A lower porosity of 0.823 was also printed for Octet topology in [79]. In this context, it is worth noting that in the direct simulations performed in [17], where the Octet topology showed clear benefits in heat transfer over the other three topologies, the Octet's strut diameter was 84 μm. In contrast, the AM parts in [79] were significantly larger with unit cell edge and the strut diameter of Octet being nearly four times and over ten times larger, respectively. Periodic unit cell simulations were further carried out by Kaur et al. [101] on TKD, rhombic dodecahedron (R-DDC), and Octet topologies, which showed that the R-DDC topology had the highest overall convective heat transfer coefficient.

The steady state experiments carried out in [79] captured the conjugate heat transfer capabilities of the 3D printed parts. For both base plate heating configurations, the Octet topology with 0.823 porosity had the highest thermal hydraulic performance, closely followed by the Cube shaped unit cell configuration. In single unit cell topologies, the endwall heat transfer plays a prominent role in the net convective transport. To address this, Kaur and Singh [69] employed transient LCT technique to determine high spatial resolution convective heat transfer coefficient at the endwalls of three unit cell topologies of the Octahedron family, viz. Octet, Octa, V-Octet (Fig. 31). As discussed earlier, this experimental procedure does not account for the conjugate heat transfer capabilities of these structures, as the experiments themselves are conducted on low thermal conductivity and low thermal diffusivity test articles. This allows the application of one-dimensional heat diffusion in a solid which can be treated as semi-infinite during the short duration experiment.

Conjugate heat transfer characterization for a larger porous media volume containing $10 \times 5 \times 5$ (L × H × W) unit cells of the same dimensions as in [79], and same four topologies (as in Fig. 30) was carried out through steady-state experiments [102]. The reported global convective heat transfer coefficient values were inclusive of the endwall effects as well as the periodic behavior of a particular unit cell topology [102]. At a given pumping power, Octet configuration provided the best overall heat transfer, closely followed by the TKD configuration [102]. In a computational study, Kaur and Singh [103] investigated the effects of surface roughness present on the struts and endwalls of the additively manufactured TKD sample $5 \times 1 \times 5$ (L × H × W) used in [79], on the local and overall heat transfer performance of the lattice frame material. This is an important

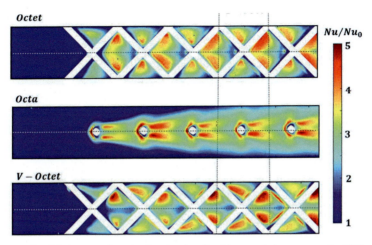

Fig. 31 Endwall heat transfer obtained using liquid crystal thermography [69].

aspect of the AM-based test article fabrication and its heat transfer characterization, since the reported global or local values of convective heat transfer are generally inclusive of the surface roughness effects, which at times, can have significant contribution. Fig. 32 shows a snapshot of the CT-scan of TKD sample additively manufactured via Binder jetting process, showcasing typical roughness levels (26 μm) observed at the endwall and computationally predicted local convective heat transfer coefficient on a smooth configuration in contrast with a 3D reconstructed geometry based on the raw CT-scan data.

4.2 Forced convection with some other working fluids (water, hydrocarbons, particles, supercritical carbon dioxide)

The investigation of flow and thermal transport in architectured cellular materials has extended to a variety of working fluids such as water, hydrocarbons, particles, sCO_2 among others. Tetrahedral lattice in sandwich configuration was explored in [104] where water was used as the working fluid. Note that thermal conductivity of water is ~25 times greater than that of air and its Prandtl number is ~10 times greater than of air. These substantial differences contribute to an order of magnitude cooling performance difference with single-phase water-based cooling compared to that with air. Kaur and Singh [105] carried out direct simulations for laminar flow of water through Octet-unit cell-based sandwich configuration which featured five interconnected unit cells in the streamwise direction. Fig. 33 shows the local convective heat

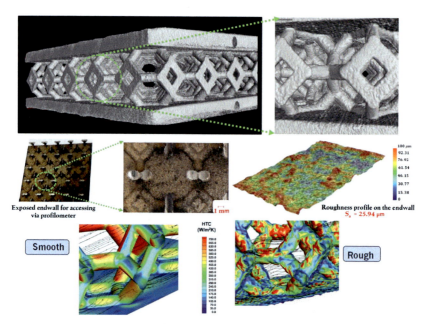

Fig. 32 CT-scanning of TKD lattice used in [79], typical roughness levels found at the endwall, and computational predictions of convective heat transfer coefficient at the struts and endwalls of a smooth configuration in comparison to its rough counterpart [103].

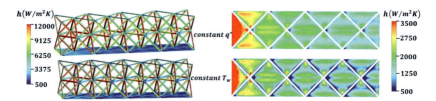

Fig. 33 Local convective heat transfer coefficient at the struts and endwall for constant heat flux and constant wall temperature boundary conditions [105].

transfer coefficient at the struts and the endwalls. Consistent with the findings in their previous work involving air as the working fluid [17], the Octet topology offers multiple stagnation-based enhanced heat transfer regions for water-based convection, along with the reduction of low heat transfer zones (typically expected on rear end of the struts) through keeping the strut diameter small. Direct simulations predicted that single-phase convective heat transfer coefficient at the struts can reach as high as $10\,\mathrm{kW/m^2K}$ for flow velocity of $0.1\,\mathrm{m/s}$, which is an order of magnitude higher than air-based convection.

Kaur and Singh [106] further extended the investigation on water-based cooling solutions, by including TKD and R-DDC unit cell topologies alongside the Octet. They demonstrated that for laminar flow through TKD, R-DDC and Octet, the R-DDC configuration had the highest thermal hydraulic performance when accounting for the product of convective heat transfer and the wetted surface area.

In the realm of AM, a key advantage lies in the ability to achieve functionalized grading of the resultant porous media. Particularly, for internal flows, with one or two walls subjected to heat load, the near-wall porosity can be increased locally to increase the flow velocities. In an experimental investigation conducted by Tseng et al. [107], they exploited this feature and demonstrated that functionalized graded samples reduced the thermal resistance by 26% compared to uniform porosity, at a fixed pumping power. Moreover, the design freedom afforded by AM enables the pursuit of multi-objective optimization of architectured cellular materials. In a noteworthy study, Takezawa et al. [108] conducted an optimization study to obtain a 3D distribution of the porosity. This study considered not only the default aspects of flow and thermal transport, but also factor in linear elasticity.

The aforementioned studies employed water as working fluid with architectured cellular materials to augment heat transfer with the substrate. In thermal management applications, the use of water spans across low temperature heat exchangers, where aluminum or copper-based AM designs can be explored to fully utilize the solid intrinsic thermal conductivity while capitalizing on the enhancement provided by high performance additively manufactured porous media. Recent advancements in material science and engineering research pertaining to AM have recently focused on the development of high-temperature materials and high-performance cooling concepts for applications such as aerothermal heating in hypersonic flights, concentrating solar thermal power, and other heat exchangers that operate at temperatures exceeding $600\,^\circ\mathrm{C}$. For aerospace application, a computational study was conducted on a pyramidal lattice in a sandwich configuration by Song et al. [109] for regenerative cooling with n-decane as working fluid for scramjets. The computational findings demonstrated a remarkable 35% reduction in wall temperature at a lattice porosity of 0.89, with pyramidal configuration in comparison to regular smooth channels.

For nuclear and solar energy areas, supercritical carbon dioxide (sCO_2) has been actively investigated due to its superior heat transfer characteristics near the pseudo-critical point [110]. Operation at near this point, where thermophysical properties vary sharply, however, introduces challenges such as "Heat Transfer Deterioration" (HTD). This phenomenon, characterized by a sudden increase in local wall temperature arising from lower convective heat transfer coefficient due to a sudden increase in the heat flux or reduction in sCO_2 mass flux, can potentially lead to component failure due to large local variation in thermal stresses. To address this, lattice frame materials have been employed. Yang et al. [111] computationally studied the benefits of employing BCC unit cells in a tube, and demonstrated that the peak temperature levels can be reduced as much as 70 K. In a follow-up investigation [112], the group conducted experiments to characterize the local convective heat transfer for sCO2 flow in a circular tube featuring BCC lattice structures. Fig. 34 shows the effect of mass flux at a fixed wall heat flux and the effect of heat flux at a fixed mass flux. The convection heat transfer coefficient reached upwards of 10 kW/m²K for a mass flux of 419 kg/m²s. Further, with increasing heat flux, the convection heat transfer coefficient exhibited a rapid rise for lower heat flux values, with this rate of increase diminishing with increasing heat flux.

The HTD issue can potentially occur in the regenerative cooling systems of scramjets (Fig. 35A), where hydrocarbons' extremely effective heat transfer properties are utilized through near critical point operation (Fig. 35B shows thermophysical property variation for n-decane). Addressing this HTD concern in hydrocarbon-based regenerative cooling, Kaur and Singh [115] recently proposed the periodic placement of TKD

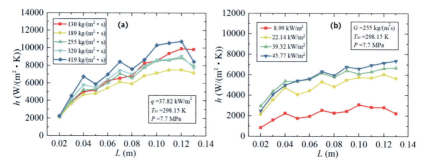

Fig. 34 Local convective heat transfer coefficient in a circular tube featuring BCC lattices, (A) effect of mass flux at a fixed heat flux, (B) effect of heat flux at a fixed mass flux [112].

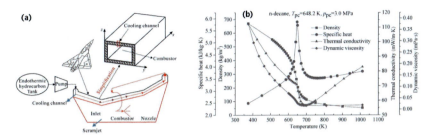

Fig. 35 (A) Typical regenerative cooling system in a scramjet [113], (B) thermophysical property variation of n-decane around the critical temperature and pressure levels [114].

unit cell in a typical minichannel designed for the combustor cooling of a scramjet. This innovative concept based on the strategic placement of discrete TKD unit cells, resulted in a peak temperature reduction of ~200 K, and an additional advantage of the suppression of wall temperature increase. Furthermore, the local convective heat transfer coefficient was enhanced by over 100% compared to a geometrically similar smooth channel operating at similar inlet conditions for the n-decane working fluid.

The application of architectured cellular materials is not limited to fluids which can flow through the complex pores and interact with the interconnected struts, leading to enhanced convective heat transfer. Recently, Aider et al. [80] demonstrated superior convective transport for a packed bed of spherical particles moving under the influence of gravity, through an array of Octet unit cells. This work addressed the challenge of low convective heat transfer on the particle side, particularly, in a typical particle-to-sCO$_2$ heat exchanger for concentrating solar power plant, which features moving packed bed between parallel plates [116,117]. In this application, particles serve both as the thermal energy storage media during transit through the receiver section, and as the "working fluid" within the shell-and-plate heat exchanger, where sCO$_2$ serves as the cold fluid. The inherently low thermal conductivity of silica particles, and poor contact between moving packed bed of particles and the parallel plates of shell-and-plate heat exchangers contribute to low levels of heat transfer. This limitation becomes a bottleneck towards the realization of levelized cost of electricity production targets of $0.05/kWh through solar energy. Aider et al. [80] leveraged the multifunctional attributes of Octet lattice, which not only provides high effective thermal conductivity and a high interstitial

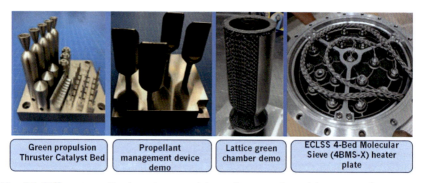

Fig. 36 Different applications in propulsion where conformal lattice configurations can provide efficient thermal management *(Source: NASA)*.

heat transfer coefficient for the particle side, but also provides high structural integrity to the shell-and-plate assembly. The Octet lattice array configured in a sandwich structure representing a single unit cell of a larger shell-and-plate heat exchanger, yielded over 40% increase in convective heat transfer compared to a geometrically similar parallel plate configuration (absence of lattice). Further investigation by Kaur et al. [81] demonstrated the enhanced heat transfer mechanism of granular flow through a similar Octet unit cell based configuration using discrete element modeling.

The highlighted applications of architectured cellular materials discussed above underscore their remarkable versatility, indicating that the current exploration is just the tip of the iceberg. The trajectory of ongoing investigations is expected to expand significantly as AM technology matures. With the evolution of this technology, there arises the potential to fabricate even more complex and finely detailed topologies, scaling down to smaller- dimensions with precision and seamless integration.

4.3 Architectured cellular materials for applications requiring conformal cooling solutions

The above studies on architectured materials were primarily based on channel flows with rectangular or square-shaped he cross-section. There are, however, many applications where such simplified flow channels do not exist, while the cooling demands require advanced concepts to be employed in such geometries. Previous demonstrations of the applicability of stochastic foams in circular channels have shown promise. However, challenges persist, particularly in the machining of metallic foams, which remains both difficult and expensive (Fig. 36).

Furthermore, the issue of bonding of custom foam configurations in complex flow channels needs to be addressed. In pursuit of versatile solutions, AM–based conformal lattice configurations are being actively explored, particularly in developing integrated cooling solutions to tackle complex thermal management challenges. The National Aeronautics and Space Administration Marshall Space Flight Center has explored different lattice unit cell topologies for a wide range of thermal management applications, as shown in Fig. 33.

A particularly impactful domain where conformal cooling solutions hold great promise is that of circular and annular flow channels. Many heat exchangers involve circular tubes to transport working fluids, typically in smooth channels. However, the integration of architectured cellular materials within a connected circular tube offers a transformative approach to enhance convective heat transfer, as exemplified in certain studies (e.g., [118]). Another noteworthy application where annular section is used for thermal management is that of a gas turbine combustor, where coolant is routed in an annular chamber to externally cool the liner walls subjected to high heat loads. The conventional approach to enhance the cooling performance of the backside of combustor liner walls, involves the use of impingement-effusion systems [119], which further complicates the design by the addition of an extra cylinder. Such expensive cooling modifications have been considered largely unavoidable, mainly due to the limitations of existing cooling concepts such as jet impingement, which entails drilling of holes for coolant routing. In addressing these challenges, conformal additively manufactured cellular material-based concepts can significantly improve the cooling efficiency while simultaneously reducing fabrication and maintenance costs.

5. Concluding remarks

The synergistic combination of advanced simulation techniques and AM has opened up new avenues for customizing cellular materials tailored to specific applications, pushing the boundaries of what is achievable in terms of performance and functionality. This article provides a comprehensive overview of these recent developments in architectured cellular materials. To provide context to these advancements, we first provide an overview of earlier research on high porosity stochastic metal foams. This research involved experimental determination of flow and thermal transport quantities, such as permeability, inertial coefficient, effective thermal conductivity, and interstitial heat transfer coefficients. Semi-empirical relationships were developed based on simplified two-dimensional assumptions, frequently used in VAR computations which were able to capture the bulk thermal transport characteristics in simple flow geometries.

The landscape of research in porous media-based enhanced heat transfer has undergone a significant transformation over the past two decades. This transformation is attributable to various factors, such as the development of high spatial resolution non-intrusive diagnostic tools, advancements in computing power, and, notably, the evolution of AM science and technology. Conventional VAR computations have given way to pore-scale, high spatial resolution direct simulations conducted on accurate 3D reconstructed models of stochastic foams. This shift has significantly improved our fundamental understanding of fluid flow and heat transfer within stochastic foams produced through the foaming process. The integration of this approach with state-of-the-art AM techniques has enabled the engineering of novel and multi-functional unit cell topologies. These topologies are tunable at both the pore- and strut-levels, and the benefits of such modifications can be realized on a larger physical scale by reticulating the unit cells in three dimensions and additively manufacturing them to the design specifications.

Over the past decade, several high-performance strut-based lattice topologies have been developed for a wide range of applications, ranging from electronics cooling and gas turbine blade cooling to propulsion, hypersonic systems, solar applications, and more. Future research on cellular materials is expected to leverage the ongoing advancements in AM and material science, specifically related to high temperature materials which have superior mechanical properties as well as high thermal conductivity. The advancements in high temperature alloys and ceramics will be crucial in the realization of advanced heat exchanger concepts, which have been recently developed either through small-scale laboratory experiments or through pore-scale computations. In addition, we anticipate investigations on bio-inspired cellular structures and smart materials with tunable thermal properties. The authors hope that this article will serve as a platform and motivation for the academic and professional community, encouraging further advancements in the exciting field of architectured cellular materials for diverse applications. The potential for transformative impact is vast, encompassing a broad spectrum of fields.

References

[1] M.S. Phanikumar, R.L. Mahajan, Non-Darcy natural convection in high porosity metal foams, Int. J. Heat. Mass. Transf. 45 (2002) 3781–3793, https://doi.org/10.1016/S0017-9310(02)00089-3

[2] Application of Foam Metal Technology to Aircraft Systems-Direction and Status. https://apps.dtic.mil/sti/citations/ADA400826, n.d. Accessed 11.11.23.

[3] X.-H. Han, Q. Wang, Y.-G. Park, C. T'Joen, A. Sommers, A. Jacobi, A review of metal foam and metal matrix composites for heat exchangers and heat sinks, Heat. Transf. Eng. 33 (2012) 991–1009, https://doi.org/10.1080/01457632.2012.659613

[4] J.Y. Ho, K.C. Leong, T.N. Wong, Additively-manufactured metallic porous lattice heat exchangers for air-side heat transfer enhancement, Int. J. Heat. Mass. Transf. 150 (2020) 119262, https://doi.org/10.1016/j.ijheatmasstransfer.2019.119262

[5] K.J. Maloney, K.D. Fink, T.A. Schaedler, J.A. Kolodziejska, A.J. Jacobsen, C.S. Roper, Multifunctional heat exchangers derived from three-dimensional micro-lattice structures, Int. J. Heat. Mass. Transf. 55 (2012) 2486–2493, https://doi.org/10.1016/j.ijheatmasstransfer.2012.01.011

[6] L.M. Niebylski, R.J. Fanning, Metal foams as energy absorbers for automobile bumpers, SAE Trans. 81 (1972) 1676–1682.

[7] J. Banhart, J. Baumeister, Deformation characteristics of metal foams, J. Mater. Sci. 33 (1998) 1431–1440, https://doi.org/10.1023/A:1004383222228

[8] Z. Ozdemir, E. Hernandez-Nava, A. Tyas, J.A. Warren, S.D. Fay, R. Goodall, et al., Energy absorption in lattice structures in dynamics: experiments, Int. J. Impact Eng. 89 (2016) 49–61, https://doi.org/10.1016/j.ijimpeng.2015.10.007

[9] L. Bai, C. Gong, X. Chen, Y. Sun, L. Xin, H. Pu, et al., Mechanical properties and energy absorption capabilities of functionally graded lattice structures: experiments and simulations, Int. J. Mech. Sci. 182 (2020) 105735, https://doi.org/10.1016/j.ijmecsci.2020.105735

[10] H. Liu, J. Wei, Z. Qu, Prediction of aerodynamic noise reduction by using open-cell metal foam, J. Sound. Vib. 331 (2012) 1483–1497, https://doi.org/10.1016/j.jsv.2011.11.016

[11] C. Teruna, F. Manegar, F. Avallone, D. Ragni, D. Casalino, T. Carolus, Noise reduction mechanisms of an open-cell metal-foam trailing edge, J. Fluid Mech. 898 (2020) A18, https://doi.org/10.1017/jfm.2020.363

[12] H. Ronge, S. Krishnan, S. Ramamoorthy, Evaluation of stochastic and periodic cellular materials for combined heat dissipation and noise reduction: experiments and modeling, IEEE Trans. Compon. Packag. Manuf. Technol. 10 (2020) 1185–1203, https://doi.org/10.1109/TCPMT.2020.2998078

[13] M. Guden, E. Celik, S. Cetiner, A. Aydin, Metals foams for biomedical applications: processing and mechanical properties, in: N. Hasirci, V. Hasirci (Eds.), Biomaterials, Springer US, Boston, MA, 2004, pp. 257–266. https://doi.org/10.1007/978-0-306-48584-8_20.

[14] L.E. Murr, S.M. Gaytan, F. Medina, H. Lopez, E. Martinez, B.I. Machado, et al., Next-generation biomedical implants using additive manufacturing of complex, cellular and functional mesh arrays, Philos. Trans. R. Soc. Math. Phys. Eng. Sci. 368 (2010) 1999–2032, https://doi.org/10.1098/rsta.2010.0010

[15] C.Y. Zhao, Review on thermal transport in high porosity cellular metal foams with open cells, Int. J. Heat. Mass. Transf. 55 (2012) 3618–3632, https://doi.org/10.1016/j.ijheatmasstransfer.2012.03.017

[16] C.Y. Zhao, T. Kim, T.J. Lu, H.P. Hodson, Thermal transport in high porosity cellular metal foams, J. Thermophys. Heat. Transf. 18 (2004) 309–317, https://doi.org/10.2514/1.11780

[17] I. Kaur, P. Singh, Flow and thermal transport through unit cell topologies of cubic and octahedron families, Int. J. Heat. Mass. Transf. 158 (2020) 119784, https://doi.org/10.1016/j.ijhcatmasstransfer.2020.119784

[18] A. Bhattacharya, V.V. Calmidi, R.L. Mahajan, Thermophysical properties of high porosity metal foams, Int. J. Heat. Mass. Transf. 45 (2002) 1017–1031, https://doi.org/10.1016/S0017-9310(01)00220-4

[19] V.V. Calmidi, R.L. Mahajan, The effective thermal conductivity of high porosity fibrous metal foams, J. Heat. Transf. 121 (1999) 466–471, https://doi.org/10.1115/1.2826001

[20] P. Singh, K. Nithyanandam, M. Zhang, R.L. Mahajan, The effect of metal foam thickness on jet array impingement heat transfer in high-porosity aluminum foams, J. Heat. Transf. 142 (2020), https://doi.org/10.1115/1.4045640

[21] K. Boomsma, D. Poulikakos, On the effective thermal conductivity of a three-dimensionally structured fluid-saturated metal foam, Int. J. Heat. Mass. Transf. 44 (2001) 827–836, https://doi.org/10.1016/S0017-9310(00)00123-X

[22] Z. Dai, K. Nawaz, Y.G. Park, J. Bock, A.M. Jacobi, Correcting and extending the Boomsma–Poulikakos effective thermal conductivity model for three-dimensional, fluid-saturated metal foams, Int. Commun. Heat. Mass. Transf. 37 (2010) 575–580, https://doi.org/10.1016/j.icheatmasstransfer.2010.01.015

[23] V.V. Calmidi, R.L. Mahajan, Forced convection in high porosity metal foams, J. Heat. Transf. 122 (2000) 557–565, https://doi.org/10.1115/1.1287793

[24] A. Bhattacharya, R.L. Mahajan, Metal foam and finned metal foam heat sinks for electronics cooling in buoyancy-induced convection, J. Electron. Packag. 128 (2005) 259–266, https://doi.org/10.1115/1.2229225

[25] A. Bhattacharya, R.L. Mahajan, Finned metal foam heat sinks for electronics cooling in forced convection, J. Electron. Packag. 124 (2002) 155–163, https://doi.org/10.1115/1.1464877

[26] M.L. Hunt, C.L. Tien, Effects of thermal dispersion on forced convection in fibrous media, Int. J. Heat. Mass. Transf. 31 (1988) 301–309, https://doi.org/10.1016/0017-9310(88)90013-0

[27] K. Boomsma, D. Poulikakos, F. Zwick, Metal foams as compact high performance heat exchangers, Mech. Mater. 35 (2003) 1161–1176, https://doi.org/10.1016/j.mechmat.2003.02.001

[28] P. Ranut, E. Nobile, L. Mancini, High resolution x-ray microtomography-based CFD simulation for the characterization of flow permeability and effective thermal conductivity of aluminum metal foams, Exp. Therm. Fluid Sci. 67 (2015) 30–36, https://doi.org/10.1016/j.expthermflusci.2014.10.018

[29] O. Al-Ketan, R. Rowshan, R.K. Abu Al-Rub, Topology-mechanical property relationship of 3D printed strut, skeletal, and sheet based periodic metallic cellular materials, Addit. Manuf. 19 (2018) 167–183, https://doi.org/10.1016/j.addma.2017.12.006

[30] D. Kong, Y. Zhang, S. Liu, Convective heat transfer enhancement by novel honeycomb-core in sandwich panel exchanger fabricated by additive manufacturing, Appl. Therm. Eng. 163 (2019) 114408, https://doi.org/10.1016/j.applthermaleng.2019.114408

[31] M. f Ashby, The properties of foams and lattices, Philos. Trans. R. Soc. Math. Phys. Eng. Sci. 364 (2005) 15–30, https://doi.org/10.1098/rsta.2005.1678

[32] H.N.G. Wadley, Multifunctional periodic cellular metals, Philos. Trans. R. Soc. Math. Phys. Eng. Sci. 364 (2005) 31–68, https://doi.org/10.1098/rsta.2005.1697

[33] Direct simulation of transport in open-cell metal foam, J. Heat Transfer | ASME Digital Collection. https://asmedigitalcollection.asme.org/heattransfer/article/128/8/793/477241/Direct-Simulation-of-Transport-in-Open-Cell-Metal, n.d. Accessed 11.11.23.

Thermal transport in engineered cellular materials 231

[34] K.K. Bodla, J.Y. Murthy, S.V. Garimella, Microtomography-based simulation of transport through open-cell metal foams, Numer. Heat. Transf. Part. Appl. 58 (2010) 527–544, https://doi.org/10.1080/10407782.2010.511987

[35] R. Lemlich, A theory for the limiting conductivity of polyhedral foam at low density, J. Colloid Interface Sci. 64 (1978) 107–110, https://doi.org/10.1016/0021-9797(78)90339-9

[36] M. Wang, N. Pan, Modeling and prediction of the effective thermal conductivity of random open-cell porous foams, Int. J. Heat. Mass. Transf. 51 (2008) 1325–1331, https://doi.org/10.1016/j.ijheatmasstransfer.2007.11.031

[37] P. Kumar, F. Topin, J. Vicente, Determination of effective thermal conductivity from geometrical properties: application to open cell foams, Int. J. Therm. Sci. 81 (2014) 13–28, https://doi.org/10.1016/j.ijthermalsci.2014.02.005

[38] K. Boomsma, D. Poulikakos, The effects of compression and pore size variations on the liquid flow characteristics in metal foams, J. Fluids Eng. 124 (2001) 263–272, https://doi.org/10.1115/1.1429637

[39] P. Ranut, E. Nobile, L. Mancini, High resolution microtomography-based CFD simulation of flow and heat transfer in aluminum metal foams, Appl. Therm. Eng. 69 (2014) 230–240, https://doi.org/10.1016/j.applthermaleng.2013.11.056

[40] A. Diani, K.K. Bodla, L. Rossetto, S.V. Garimella, Numerical investigation of pressure drop and heat transfer through reconstructed metal foams and comparison against experiments, Int. J. Heat. Mass. Transf. 88 (2015) 508–515, https://doi.org/10.1016/j.ijheatmasstransfer.2015.04.038

[41] P. Yu, Y. Wang, R. Ji, H. Wang, J. Bai, Pore-scale numerical study of flow characteristics in anisotropic metal foam with actual skeleton structure, Int. Commun. Heat. Mass. Transf. 126 (2021) 105401, https://doi.org/10.1016/j.icheatmasstransfer.2021.105401

[42] T. Yang, Z. Wang, H. Wang, P. Yu, Y. Wang, Numerical study of flow and heat transfer in a three-dimensional metal foam considering different direction micropores in skeleton structure, Int. Commun. Heat. Mass. Transf. 134 (2022) 106052, https://doi.org/10.1016/j.icheatmasstransfer.2022.106052

[43] A. Della Torre, G. Montenegro, G.R. Tabor, M.L. Wears, CFD characterization of flow regimes inside open cell foam substrates, Int. J. Heat. Fluid Flow. 50 (2014) 72–82, https://doi.org/10.1016/j.ijheatfluidflow.2014.05.005

[44] M. Sun, C. Hu, L. Zha, Z. Xie, L. Yang, D. Tang, et al., Pore-scale simulation of forced convection heat transfer under turbulent conditions in open-cell metal foam, Chem. Eng. J. 389 (2020) 124427, https://doi.org/10.1016/j.cej.2020.124427

[45] Z. Zhang, G. Yan, M. Sun, H. Yan, J. Zhao, Y. Song, et al., Pore-scale simulation of forced convection heat transfer in metal foams with uniform and gradient structures, Appl. Therm. Eng. 225 (2023) 120074, https://doi.org/10.1016/j.applthermaleng.2023.120074

[46] W. Sir Thomson, On the division of space with minimum partitional area, Acta Math. 11 (1887) 121–134, https://doi.org/10.1007/BF02612322

[47] D. Weaire, R. Phelan, A counter-example to Kelvin's conjecture on minimal surfaces, Philos. Mag. Lett. 69 (1994) 107–110, https://doi.org/10.1080/09500839408241577

[48] A. Kopanidis, A. Theodorakakos, E. Gavaises, D. Bouris, 3D numerical simulation of flow and conjugate heat transfer through a pore scale model of high porosity open cell metal foam, Int. J. Heat. Mass. Transf. 53 (2010) 2539–2550, https://doi.org/10.1016/j.ijheatmasstransfer.2009.12.067

[49] P. Poureslami, M. Siavashi, H. Moghimi, M. Hosseini, Pore-scale convection-conduction heat transfer and fluid flow in open-cell metal foams: a three-dimensional

multiple-relaxation time lattice Boltzmann (MRT-LBM) solution, Int. Commun. Heat. Mass. Transf. 126 (2021) 105465, https://doi.org/10.1016/j.icheatmasstransfer.2021.105465

[50] C. Moon, D. Kim, G. Bamorovat Abadi, S.Y. Yoon, K.C. Kim, Effect of ligament hollowness on heat transfer characteristics of open-cell metal foam, Int. J. Heat. Mass. Transf. 102 (2016) 911–918, https://doi.org/10.1016/j.ijheatmasstransfer.2016.06.068

[51] N.J. Dyck, A.G. Straatman, A new approach to digital generation of spherical void phase porous media microstructures, Int. J. Heat. Mass. Transf. 81 (2015) 470–477, https://doi.org/10.1016/j.ijheatmasstransfer.2014.10.017

[52] S.S. Sundarram, W. Li, The effect of pore size and porosity on thermal management performance of phase change material infiltrated microcellular metal foams, Appl. Therm. Eng. 64 (2014) 147–154, https://doi.org/10.1016/j.applthermaleng.2013.11.072

[53] R. Paknahad, M. Siavashi, M. Hosseini, Pore-scale fluid flow and conjugate heat transfer study in high porosity Voronoi metal foams using multi-relaxation-time regularized lattice Boltzmann (MRT-RLB) method, Int. Commun. Heat. Mass. Transf. 141 (2023) 106607, https://doi.org/10.1016/j.icheatmasstransfer.2022.106607

[54] Z. Nie, Y. Lin, Q. Tong, Numerical investigation of pressure drop and heat transfer through open cell foams with 3D Laguerre-Voronoi model, Int. J. Heat. Mass. Transf. 113 (2017) 819–839, https://doi.org/10.1016/j.ijheatmasstransfer.2017.05.119

[55] I. Kaur, P. Singh, Critical evaluation of additively manufactured metal lattices for viability in advanced heat exchangers, Int. J. Heat. Mass. Transf. 168 (2021) 120858, https://doi.org/10.1016/j.ijheatmasstransfer.2020.120858

[56] J. Tian, T. Kim, T.J. Lu, H.P. Hodson, D.T. Queheillalt, D.J. Sypeck, et al., The effects of topology upon fluid-flow and heat-transfer within cellular copper structures, Int. J. Heat. Mass. Transf. 47 (2004) 3171–3186, https://doi.org/10.1016/j.ijheatmasstransfer.2004.02.010

[57] J.-H. Joo, K.-J. Kang, T. Kim, T.J. Lu, Forced convective heat transfer in all metallic wire-woven bulk Kagome sandwich panels, Int. J. Heat. Mass. Transf. 54 (2011) 5658–5662, https://doi.org/10.1016/j.ijheatmasstransfer.2011.08.018

[58] U. Kemerli, K. Kahveci, Conjugate forced convective heat transfer in a sandwich panel with a Kagome truss core: the effects of strut length and diameter, Appl. Therm. Eng. 167 (2020) 114794, https://doi.org/10.1016/j.applthermaleng.2019.114794

[59] C. Hou, G. Yang, X. Wan, J. Chen, Study of thermo-fluidic characteristics for geometric-anisotropy Kagome truss-cored lattice, Chin. J. Aeronaut. 32 (2019) 1635–1645, https://doi.org/10.1016/j.cja.2019.03.023

[60] B. Shen, Y. Li, H. Yan, S.K.S. Boetcher, G. Xie, Heat transfer enhancement of wedge-shaped channels by replacing pin fins with Kagome lattice structures, Int. J. Heat. Mass. Transf. 141 (2019) 88–101, https://doi.org/10.1016/j.ijheatmasstransfer.2019.06.059

[61] X. Bai, Z. Zheng, A. Nakayama, Heat transfer performance analysis on lattice core sandwich panel structures, Int. J. Heat. Mass. Transf. 143 (2019) 118525, https://doi.org/10.1016/j.ijheatmasstransfer.2019.118525

[62] R.L. Webb, E.R.G. Eckert, Application of rough surfaces to heat exchanger design, Int. J. Heat. Mass. Transf. 15 (1972) 1647–1658, https://doi.org/10.1016/0017-9310(72)90095-6

[63] T. Kim, H.P. Hodson, T.J. Lu, Fluid-flow and endwall heat-transfer characteristics of an ultralight lattice-frame material, Int. J. Heat. Mass. Transf. 47 (2004) 1129–1140, https://doi.org/10.1016/j.ijheatmasstransfer.2003.10.012

[64] T. Kim, H.P. Hodson, T.J. Lu, Contribution of vortex structures and flow separation to local and overall pressure and heat transfer characteristics in an ultralightweight lattice material, Int. J. Heat. Mass. Transf. 48 (2005) 4243–4264, https://doi.org/10.1016/j.ijheatmasstransfer.2005.04.026

[65] P. Singh, S. Ekkad, Experimental study of heat transfer augmentation in a two-pass channel featuring V-shaped ribs and cylindrical dimples, Appl. Therm. Eng. 116 (2017) 205–216, https://doi.org/10.1016/j.applthermaleng.2017.01.098

[66] P. Singh, Y. Ji, S.V. Ekkad, Experimental and numerical investigation of heat and fluid flow in a square duct featuring criss-cross rib patterns, Appl. Therm. Eng. 128 (2018) 415–425, https://doi.org/10.1016/j.applthermaleng.2017.09.036

[67] P. Singh, Y. Ji, S.V. Ekkad, Multipass serpentine cooling designs for negating coriolis force effect on heat transfer: 45-deg angled rib turbulated channels, J. Turbomach. 141 (2019), https://doi.org/10.1115/1.4042648

[68] P. Singh, J. Pandit, S.V. Ekkad, Characterization of heat transfer enhancement and frictional losses in a two-pass square duct featuring unique combinations of rib turbulators and cylindrical dimples, Int. J. Heat. Mass. Transf. 106 (2017) 629–647, https://doi.org/10.1016/j.ijheatmasstransfer.2016.09.037

[69] I. Kaur, P. Singh, Endwall heat transfer characteristics of octahedron family lattice-frame materials, Int. Commun. Heat. Mass. Transf. 127 (2021) 105522, https://doi.org/10.1016/j.icheatmasstransfer.2021.105522

[70] S.V. Ekkad, P. Singh, Liquid crystal thermography in gas turbine heat transfer: a review on measurement techniques and recent investigations, Crystals 11 (2021) 1332, https://doi.org/10.3390/cryst11111332

[71] H.B. Yan, Q.C. Zhang, T.J. Lu, T. Kim, A lightweight x-type metallic lattice in single-phase forced convection, Int. J. Heat. Mass. Transf. 83 (2015) 273–283, https://doi.org/10.1016/j.ijheatmasstransfer.2014.11.061

[72] X. Jin, B. Shen, H. Yan, B. Sunden, G. Xie, Comparative evaluations of thermo-fluidic characteristics of sandwich panels with x-lattice and pyramidal-lattice cores, Int. J. Heat. Mass. Transf. 127 (2018) 268–282, https://doi.org/10.1016/j.ijheatmasstransfer.2018.07.087

[73] H. Yan, X. Yang, T. Lu, G. Xie, Convective heat transfer in a lightweight multi-functional sandwich panel with x-type metallic lattice core, Appl. Therm. Eng. 127 (2017) 1293–1304, https://doi.org/10.1016/j.applthermaleng.2017.08.081

[74] H.B. Yan, Q.C. Zhang, T.J. Lu, An x-type lattice cored ventilated brake disc with enhanced cooling performance, Int. J. Heat. Mass. Transf. 80 (2015) 458–468, https://doi.org/10.1016/j.ijheatmasstransfer.2014.09.060

[75] Y. Li, G. Xie, S.K.S. Boetcher, H. Yan, Heat transfer enhancement of x-lattice-cored sandwich panels by introducing pin fins, dimples or protrusions, Int. J. Heat. Mass. Transf. 141 (2019) 627–642, https://doi.org/10.1016/j.ijheatmasstransfer.2019.07.009

[76] X. Jin, Y. Li, H. Yan, G. Xie, Comparative study of flow structures and heat transfer enhancement in a metallic lattice fabricated by metal sheet folding: effects of punching location shift, Int. J. Heat. Mass. Transf. 134 (2019) 209–225, https://doi.org/10.1016/j.ijheatmasstransfer.2019.01.036

[77] M.A. Chatzopoulou, S. Lecompte, M.D. Paepe, C.N. Markides, Off-design optimisation of organic Rankine cycle (ORC) engines with different heat exchangers and volumetric expanders in waste heat recovery applications, Appl. Energy. 253 (2019) 113442, https://doi.org/10.1016/j.apenergy.2019.113442

[78] I. Kaur, P. Singh, State-of-the-art in heat exchanger additive manufacturing, Int. J. Heat. Mass. Transf. 178 (2021) 121600, https://doi.org/10.1016/j.ijheatmasstransfer.2021.121600

[79] I. Kaur, Y. Aider, K. Nithyanandam, P. Singh, Thermal-hydraulic performance of additively manufactured lattices for gas turbine blade trailing edge cooling, Appl. Therm. Eng. 211 (2022) 118461, https://doi.org/10.1016/j.applthermaleng.2022.118461

[80] Y. Aider, I. Kaur, A. Mishra, L. Li, H. Cho, J. Martinek, et al., Heat transfer characteristics of particle and air flow through additively manufactured lattice frame material based on octet-shape topology, J. Sol. Energy Eng. 145 (2023), https://doi.org/10.1115/1.4062196

[81] I. Kaur, Y. Aider, H. Cho, P. Singh, Granular flow in novel octet shape-based lattice frame material, J. Sol. Energy Eng. (2023) 1–21, https://doi.org/10.1115/1.4064018

[82] Y. Zhu, W. Peng, R. Xu, P. Jiang, Review on active thermal protection and its heat transfer for airbreathing hypersonic vehicles, Chin. J. Aeronaut. 31 (2018) 1929–1953, https://doi.org/10.1016/j.cja.2018.06.011

[83] M. Wong, I. Owen, C.J. Sutcliffe, A. Puri, Convective heat transfer and pressure losses across novel heat sinks fabricated by selective laser melting, Int. J. Heat. Mass. Transf. 52 (2009) 281–288, https://doi.org/10.1016/j.ijheatmasstransfer.2008.06.002

[84] D. Liang, W. Chen, Y. Ju, M.K. Chyu, Comparing endwall heat transfer among staggered pin fin, Kagome and body centered cubic arrays, Appl. Therm. Eng. 185 (2021) 116306, https://doi.org/10.1016/j.applthermaleng.2020.116306

[85] D. Liang, G. He, W. Chen, Y. Chen, M.K. Chyu, Fluid flow and heat transfer performance for micro-lattice structures fabricated by selective laser melting, Int. J. Therm. Sci. 172 (2022) 107312, https://doi.org/10.1016/j.ijthermalsci.2021.107312

[86] S. Parbat, Z. Min, L. Yang, M. Chyu, Experimental and numerical analysis of additively manufactured inconel 718 coupons with lattice structure, J. Turbomach. 142 (2020), https://doi.org/10.1115/1.4046527

[87] M. Shahrzadi, M. Davazdah Emami, A.H. Akbarzadeh, Heat transfer in BCC lattice materials: conduction, convection, and radiation, Compos. Struct. 284 (2022) 115159, https://doi.org/10.1016/j.compstruct.2021.115159

[88] S. Takarazawa, K. Ushijima, R. Fleischhauer, J. Kato, K. Terada, W.J. Cantwell, et al., Heat-transfer and pressure drop characteristics of micro-lattice materials fabricated by selective laser metal melting technology, Heat. Mass. Transf. 58 (2022) 125–141, https://doi.org/10.1007/s00231-021-03083-0

[89] D. Liang, W. Bai, W. Chen, M.K. Chyu, Investigating the effect of element shape of the face-centered cubic lattice structure on the flow and endwall heat transfer characteristics in a rectangular channel, Int. J. Heat. Mass. Transf. 153 (2020) 119579, https://doi.org/10.1016/j.ijheatmasstransfer.2020.119579

[90] S. Yun, J. Kwon, D. Lee, H.H. Shin, Y. Kim, Heat transfer and stress characteristics of additive manufactured FCCZ lattice channel using thermal fluid-structure interaction model, Int. J. Heat. Mass. Transf. 149 (2020) 119187, https://doi.org/10.1016/j.ijheatmasstransfer.2019.119187

[91] S. Yun, D. Lee, D.S. Jang, M. Lee, Y. Kim, Numerical analysis on thermo-fluid–structural performance of graded lattice channels produced by metal additive manufacturing, Appl. Therm. Eng. 193 (2021) 117024, https://doi.org/10.1016/j.applthermaleng.2021.117024

[92] M. Qian, J. Li, Z. Xiang, Z. Dong, J. Xiao, X. Hu, Study on heat dissipation performance of a lattice porous structures under jet impingement cooling, Case Stud. Therm. Eng. 49 (2023) 103244, https://doi.org/10.1016/j.csite.2023.103244

[93] C. Moon, H.D. Kim, K.C. Kim, Kelvin-cell-based metal foam heat exchanger with elliptical struts for low energy consumption, Appl. Therm. Eng. 144 (2018) 540–550, https://doi.org/10.1016/j.applthermaleng.2018.07.110

[94] A. Suzuki, H. Nakatani, M. Kobashi, Machine learning surrogate modeling toward the design of lattice-structured heat sinks fabricated by additive manufacturing, Mater. Des. 230 (2023) 111969, https://doi.org/10.1016/j.matdes.2023.111969

[95] J. Storm, M. Abendroth, M. Kuna, Numerical and analytical solutions for anisotropic yield surfaces of the open-cell Kelvin foam, Int. J. Mech. Sci. 105 (2016) 70–82, https://doi.org/10.1016/j.ijmecsci.2015.10.014

[96] A.K. Mishra, A. Kumar, Performance of asymmetric octet lattice structures under compressive and bending loads, Eng. Fail. Anal. 154 (2023) 107669, https://doi.org/10.1016/j.engfailanal.2023.107669

[97] A. Chaudhari, P. Ekade, S. Krishnan, Experimental investigation of heat transfer and fluid flow in octet-truss lattice geometry, Int. J. Therm. Sci. 143 (2019) 64–75, https://doi.org/10.1016/j.ijthermalsci.2019.05.003

[98] P. Ekade, S. Krishnan, Fluid flow and heat transfer characteristics of octet truss lattice geometry, Int. J. Therm. Sci. 137 (2019) 253–261, https://doi.org/10.1016/j.ijthermalsci.2018.11.031

[99] S. Krishnan, J.Y. Murthy, S.V. Garimella, Direct simulation of transport in open-cell metal foam, J. Heat. Transf. 128 (2006) 793–799, https://doi.org/10.1115/1.2227038

[100] I. Kaur, R.L. Mahajan, P. Singh, Generalized correlation for effective thermal conductivity of high porosity architectured materials and metal foams, Int. J. Heat. Mass. Transf. 200 (2023) 123512, https://doi.org/10.1016/j.ijheatmasstransfer.2022.123512

[101] I. Kaur, S. Mujahid, Y. Paudel, H. Rhee, P. Singh, Numerical analysis of flow and heat transfer characteristics of lattice-based compact heat sinks, J. Electron. Packag. 145 (2022), https://doi.org/10.1115/1.4056305

[102] Y. Aider, I. Kaur, H. Cho, P. Singh, Periodic heat transfer characteristics of additively manufactured lattices, Int. J. Heat. Mass. Transf. 189 (2022) 122692, https://doi.org/10.1016/j.ijheatmasstransfer.2022.122692

[103] I. Kaur, P. Singh, Effects of inherent surface roughness of additively manufactured lattice frame material on flow and thermal transport, Int. J. Heat. Mass. Transf. 209 (2023) 124077, https://doi.org/10.1016/j.ijheatmasstransfer.2023.124077

[104] Thermal control of composite sandwich structure with lattice truss cores, J. Thermophys. Heat Transfer. https://arc.aiaa.org/doi/full/10.2514/1.T4361, (n.d.) Accessed 24.11.23.

[105] I. Kaur, P. Singh, Numerical investigation on conjugate heat transfer in octet-shape-based single unit cell thick metal foam, Int. Commun. Heat. Mass. Transf. 121 (2021) 105090, https://doi.org/10.1016/j.icheatmasstransfer.2020.105090

[106] I. Kaur, P. Singh, Conjugate heat transfer in lattice frame materials based on novel unit cell topologies, Numer. Heat. Transf. Part. Appl. 82 (2022) 788–801, https://doi.org/10.1080/10407782.2022.2083874

[107] P.-H. Tseng, K.-T. Tsai, A.-L. Chen, C.-C. Wang, Performance of novel liquid-cooled porous heat sink via 3-D laser additive manufacturing, Int. J. Heat. Mass. Transf. 137 (2019) 558–564, https://doi.org/10.1016/j.ijheatmasstransfer.2019.03.116

[108] Optimization of an additively manufactured functionally graded lattice structure with liquid cooling considering structural performances, ScienceDirect https://www.sciencedirect.com/science/article/pii/S0017931019315066, n.d. Accessed 24.11.23.

[109] J. Song, Y. Yuan, J. Liu, S. Luo, B. Sunden, Heat transfer enhancement of regenerative cooling channel with pyramid lattice sandwich structures, Heat. Transf. Eng. 44 (2023) 1271–1285, https://doi.org/10.1080/01457632.2022.2127049

[110] D.E. Kim, M.H. Kim, Experimental study of the effects of flow acceleration and buoyancy on heat transfer in a supercritical fluid flow in a circular tube, Nucl. Eng. Des. 240 (2010) 3336–3349, https://doi.org/10.1016/j.nucengdes.2010.07.002

[111] Z. Yang, X. Luo, W. Chen, M.K. Chyu, Mitigation effects of body-centered cubic lattices on the heat transfer deterioration of supercritical CO_2, Appl. Therm. Eng. 183 (2021) 116085, https://doi.org/10.1016/j.applthermaleng.2020.116085

[112] X. Shi, Z. Yang, W. Chen, M.K. Chyu, Investigation of the effect of lattice structure on the fluid flow and heat transfer of supercritical CO_2 in tubes, Appl. Therm. Eng. 207 (2022) 118132, https://doi.org/10.1016/j.applthermaleng.2022.118132

[113] Y. Li, F. Sun, G. Xie, J. Qin, Improved thermal performance of cooling channels with truncated ribs for a scramjet combustor fueled by endothermic hydrocarbon, Appl. Therm. Eng. 142 (2018) 695–708, https://doi.org/10.1016/j.applthermaleng.2018.07.055

[114] Y. Chen, Y. Li, B. Sunden, G. Xie, The abnormal heat transfer behavior of super-critical n–decane flowing in a horizontal tube under regenerative cooling for scramjet engines, Appl. Therm. Eng. 167 (2020) 114637, https://doi.org/10.1016/j.applthermaleng.2019.114637

[115] I. Kaur, P. Singh, Flow and thermal transport of supercritical n–decane in square minichannel featuring uniformly spaced tetrakaidecahedron-shaped unit cells, Int. Commun. Heat. Mass. Transf. 145 (2023) 106835, https://doi.org/10.1016/j.icheatmasstransfer.2023.106835

[116] K.J. Albrecht, C.K. Ho, Design and operating considerations for a shell-and-plate, moving packed-bed, particle-to-sCO2 heat exchanger, Sol. Energy. 178 (2019) 331–340, https://doi.org/10.1016/j.solener.2018.11.065

[117] K.J. Albrecht, C.K. Ho, Heat transfer models of moving packed-bed particle-to-sCO2 heat exchangers, J. Sol. Energy Eng. 141 (2018), https://doi.org/10.1115/1.4041546

[118] J.Y. Ho, K.C. Leong, Cylindrical porous inserts for enhancing the thermal and hydraulic performance of water-cooled cold plates, Appl. Therm. Eng. 121 (2017) 863–878, https://doi.org/10.1016/j.applthermaleng.2017.04.101

[119] B. Wurm, A. Schulz, H.-J. Bauer, M. Gerendas, Impact of swirl flow on the cooling performance of an effusion cooled combustor liner, J. Eng. Gas. Turbines Power. 134 (2012), https://doi.org/10.1115/1.4007332

CHAPTER SIX

Heat transfer analysis of partially ionized hybrid nanofluids flow comprising magnetic/non-magnetic nanoparticles in an annular region of two homocentric inclined cylinders

Muhammad Ramzan* and Nazia Shahmir
Department of Computer Science, Bahria University, Islamabad, Pakistan
*Corresponding author. e-mail address: mramzan@bahria.edu.pk

Contents

1. Introduction	238
2. Mathematical formulation	241
2.1 Physical quantities	244
2.2 Numerical analysis	245
2.3 Outcomes with discussion	245
2.4 Future areas of research	251
3. Conclusions	252
References	252

Abstract

This work analyzes the flow of magnetic and non-magnetic nanoparticles in hybrid nanofluids in an annular region of two homocentric slanted cylinders influenced by Hall currents and ion slip effect with linear thermal convection in a porous medium. Silver and copper are used as non-magnetic nanoparticles whereas cobalt ferrite and magnesium zinc ferrite are taken as magnetic nanoparticles immersed in aqueous-based hybrid nanofluids. Heat transmission analysis is performed considering the effects of viscous/ohmic dissipation along with Cattaneo Christov heat flux. At the cylinder's walls, temperature jump and velocity slip conditions are assumed. Moreover, a temperature-dependent thermal conductivity model is engaged. Numerical results are obtained via the bvp4c scheme and are presented in the form of graphs and tables. Results revealed that for greater values of the Hartmann number, the velocity distribution dwindled significantly for hybrid nanofluid flow containing magnetic nanoparticles. Moreover, the heat transmission rate is more pronounced for hybrid nanofluid-containing magnetic nanoparticles.

Advances in Heat Transfer, Volume 57
ISSN 0065-2717, https://doi.org/10.1016/bs.aiht.2024.04.001
Copyright © 2024 Elsevier Inc. All rights are reserved, including for text and data mining, AI training, and similar technologies.

Nomenclature

δ	Channel width
g	Gravity
H_a	Hartman number
γ	Angle of inclined concentric cylinders
R	Dimensionless radial coordinate
ρ_{HNF}	Density of hybrid nanofluid
R_e	Reynolds number
δ	Distance between the plates
κ	Variable thermal conductivity
σ_t	Tangential momentum accommodation
Pr	Prandtl number
λ_{cc}	Thermal relaxation parameter
Υ	Wall-fluid interaction
T_1	Temperature at the outer wall
$g(\eta)$	Dimensionless secondary velocity
$(\beta_1)_{HNF}$	Thermal expansion coefficient
U	Dimensionless axial velocity
k_{HNF}	Thermal conductivity of hybrid nanofluid
D_c	Diffusivity of microorganisms
ϖ	Angle of applied magnetic field
Φ_{HNF}	Particles volume fraction
$\Theta(R)$	Dimensionless temperature
β_0	Magnetic field strength
γ_c	Thermal relaxation coefficient
$U(R)$	Dimensionless axial velocity
r_1, r_2	Inner and outer cylinders radii
N_u	Nusselt number
ε	Porosity parameter
K_n	Knudsen number
T_o	Temperature at the inner wall
s	Radii ratio parameter
B_r	Brinkman number

1. Introduction

In today's era, nano, and hybrid nanotechnologies have garnered significant attention from academics, researchers, and scientists, particularly in the realm of flow and heat transfer processes. This is largely due to their extensive applicability in engineering and industrial domains. The term "nanofluid" was coined by Choi [1,2], referring to fluids composed of a liquid base phase (such as ethylene glycol, toluene, water, etc.) with a suspended solid phase (including metals, metal oxides, metal carbides, etc.). It's crucial to precisely determine the volume concentration of nanoparticles, ensuring minimal impact on fluid flow characteristics. These nanofluids find utility as coolants in various equipment like microelectronic devices, heat exchangers, radiators, etc., where efficient thermal transport is essential. Nevertheless, there's a pressing need to enhance rheological properties, particularly thermal conductivity. This necessity has led to the emergence of hybrid nanofluids, which exhibit superior thermal performance compared to those containing single nanoparticles. Hybrid nanofluids boast a broad spectrum of applications, including refrigeration, ventilation, air conditioning, heat pipes, coolant systems in machinery and industrial processes, electronic device cooling, aerospace and automotive industries, power stations, nuclear plant cooling, and biomedical applications. Shaheen et al. [3] emphasized the flow properties of a tri-hybrid nanofluid over an expanding or shrinking cylinder with Thompson and Troian slip conditions. The investigation incorporates surface-catalyzed reactions and a modified Fourier law. The findings indicate an increase in the surface drag coefficient due to porosity and curvature parameters, leading to improved thermal performance against Newtonian heating effects. In a related study, Shahmir et al. [4] compared the flows of nano and hybrid nanofluids over an elongated cylinder within a spongy medium. The base fluid for both types of nanofluids is a mixture of water and ethylene glycol in a 50:50 ratio. Additionally, the study examines the shapes of nanoparticles. Results show a 0.83% enhancement in heat transfer efficiency for the hybrid nanofluid compared to the simple nanofluid configuration. In another contemporary investigation, Gul et al. [5] conducted a comparative investigation into the heat transfer efficiency between hybrid and tri-hybrid nanofluid flows over a cylindrical surface. This study incorporates considerations for homogeneous-heterogeneous reactions alongside the Darcy-Forchheimer effects. The analysis also takes into account induced magnetic field effects and considers the slip condition at the surface of the cylinder. Three different base fluids—water, engine oil, and kerosene oil—are employed. The findings reveal a 19.14% higher heat transfer rate for engine oil compared to kerosene oil and a 12.83% increase compared to water. The flow of the nano/hybrid nanofluid flow between coaxial cylinders influenced by magnetohydrodynamics is numerically assessed by Alsaedi et al. [6]. The nanofluid is comprised of Cu/Kerosene oil and the hybrid nanofluid contains GO + Cu/Kerosene oil combination. The inner cylinder is considered fixed while the outer is assumed to be rotating. Both the hybrid and nanofluids show a decline in the fluid temperature for the magnetohydrodynamics effect. Tayebi and Chamkha [7] investigated the flow characteristics of hybrid and nanofluids within an annular space formed by confocal elliptical cylinders. In this setup, the inner cylinder is heated while the outer cylinder has isothermal walls. An intriguing finding revealed that the heat transfer rate is more efficient in the case of the hybrid fluid compared to the nanofluid. Some more recent attempts highlighting the flows of the hybrid nanofluids considering diverse configurations may be found in [8–14].

The role of the Hall current and ion slip in plasma physics can't be ruled out owing to its wide applicability in a variety of scenarios like space physics, sensor technology, and nuclear fusion research. Hall current and ion slip phenomena occur when electrons and ions are in a diverse and coherent position respectively. Scientists and researchers have demonstrated keen interest in studying the Hall current and ion slip effects in numerous scenarios. Iqbal et al. [15] outlined the partially ionized Eyring Powell nanofluid flow under the influence of motile organisms and Cattaneo-Christov heat flux engaged a semi–analytical Homotopy analysis method. It is observed here that a strong magnetic field hinders the flow of the fluid. In another study, considering the impact of the Hall current and ion slip, Salmi et al. [16] numerically discussed the Williamson hybrid nanofluid flow in a permeable medium. The model incorporates the double diffusion Cattaneo-Christov heat flux and discusses the impact of thermal relaxation and retardation times. Moreover, the process of ohmic dissipation diminishes with the enhanced effects of the partially ionized fluid. Ramzan et al. [17] delineated the flow of partially ionized Casson nanofluid within an annular region between two concentric cylinders, accounting for the influence of motile organisms and Joule heating. They employed the Buongiorno nanoscale model to address the nanofluid's thermophoretic and Brownian motion effects. The fluid temperature experiences enhancement with improved Joule heating. Furthermore, Patil and Shankar [18] numerically examined the flow of an aqueous-based hybrid nanofluid around a yawed cylinder, revealing a reduction in fluid velocity due to the surface drag coefficient. Additionally, Chung et al. [19] numerically discussed the flow of a hybrid nanofluid over a three-dimensional surface extended nonlinearly, considering a non-uniform source/sink with thermal stratification. They observed an increase in the rate of heat flux against the thermal stratification effect.

The review of existing literature reveals that there have been efforts to analyze the flow characteristics of hybrid nanofluids, incorporating

Hall current and ion slip effects. However, there remains a notable gap in research focusing on hybrid nanofluid flows within the annular region between concentric cylinders. To date, no study has investigated the flow behavior of partially ionized aqueous-based hybrid nanofluids containing magnetic or non-magnetic nanoparticles within such a configuration. An analysis of heat transfer is conducted, taking into account viscous and ohmic dissipations, along with the Cattaneo-Christov heat flux. Temperature jump and velocity slip conditions are assumed at the walls of the cylinders, and a model with temperature-dependent thermal conductivity is employed. The numerical results are presented in the form of graphs and tables for analysis.

2. Mathematical formulation

Let us consider the role of Hall and ion slip on the magnetic and non-magnetic nanoparticles in the hybrid nanofluid in an inclined concentric cylinder with an angle γ as depicted in Fig. 1. The influence of buoyancy force is induced by the temperature difference between the cylinders and in the vicinity's nanofluid, which plays the role of g. Further, the model is based on a two-dimensional coordinate system with the

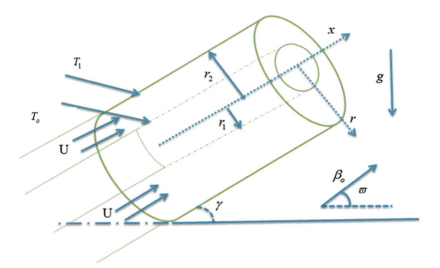

Fig. 1 Geometrical illustration.

Table 1 Thermal and physical values for magnetic/non-magnetic nanoparticles and water.

	Density	Thermal conductivity	Specific heat	Thermal expansion	Electrical conductivity
(Ag)	10490	429	235	1.89×10^{-6}	6.30×10^{7}
(Cu)	8933	401	385	1.67×10^{-6}	5.96×10^{7}
$(CoFe_2O_4)$	4907	3.5	700	3.9×10^{-6}	3.9×10^{-6}
$(Mn - ZnFe_2O_4)$	4900	5	800	8.0×10^{-6}	8×10^{-6}
(H_2O)	997.1	0.613	4179	21×10^{-5}	0.05

z-axis is aligned in a vertical direction and the r-Axis axis is taken perpendicular to the walls of the cylinders. Next, r_1 illustrates the radius at the inner cylindrical surface with temperature T_0, and r_2 denote the radius at the outer cylindrical surface with temperature T_1.

Table 1 is constructed to illustrate the thermal and physical characteristics of both the conventional fluid and the hypothetical nanoparticles.

Considering the assumptions stated above, the governing system of equations can be formulated as follows:

$$\frac{1}{r}\frac{d}{dr}(ru) = 0,$$ (1)

$$-\frac{dp}{dr} = \frac{\mu_{HNF}}{\rho_{HNF}}\left[\frac{1}{r}\frac{d}{dr}\left(r\frac{du}{dr}\right)\right] + g\cos\gamma\left[(\rho\beta_1)_{HNF}(T - T_0)\right] - \frac{\mu_{HNF}}{\rho_{HNF}\bar{k}}u -$$
$$\frac{\sigma_{HNF}\beta_0^2}{\rho_{HNF}}\left[\frac{\cos^2\varpi\,(\beta_e\beta_i + 1)}{(\beta_e\beta_i + 1)(1 + \beta_e\beta_i\cos^2\varpi) + \beta_e^2\cos^2\varpi}\right]u = 0,$$ (2)

$$\frac{1}{(\rho C_p)_{HNF}}\frac{1}{r}\left[\frac{d}{dr}\left(k_{HNF}(T)\,r\frac{dT}{dr}\right)\right] + \frac{\nu_{HNF}}{(\rho C_p)_{HNF}}\left(\frac{\partial u}{\partial r}\right)^2$$
$$- \gamma_c\left(u\frac{du}{dr}\frac{dT}{dr} + u^2\frac{d^2T}{dr^2}\right) + \left(\frac{\sigma_{HNF}\beta_0^2}{\rho_{HNF}}\right)u^2 = 0,$$ (3)

Heat transfer analysis

With boundary constraints:

$$u(r) = \left[\frac{\sqrt{\pi R T_0 / 2\mu_F}}{\rho_F} \frac{(2 - \sigma_v)}{(\sigma_v)} \right] \frac{du}{dr} \Bigg|_{r=r_1},$$

$$u(r) = -\left[\frac{\sqrt{\pi R T_0 / 2\mu_F}}{\rho_F} \frac{(2 - \sigma_v)}{(\sigma_v)} \right] \frac{du}{dr} \Bigg|_{r=r_2},$$

$$T(r) = T_1 + \left[\frac{2\gamma^*(2 - \sigma_t)}{(\gamma^* + 1)\sigma_t} \frac{\sqrt{\pi R T_0 / 2\mu_F}}{\rho_F} \right] \left(\frac{1}{\mathrm{Pr}} \right) \frac{dT}{dr} \Bigg|_{r=r_1},$$

$$T(r) = T_0 - \left[\frac{2\gamma^*(2 - \sigma_t)}{(\gamma^* + 1)\sigma_t} \frac{\sqrt{\pi R T_0 / 2\mu_F}}{\rho_F} \right] \left(\frac{1}{\mathrm{Pr}} \right) \frac{dT}{dr} \Bigg|_{r=r_2}.$$

(4)

With appropriate constraints:

$$R = \frac{r - r_1}{\delta}, \; \delta = r_2 - r_1, \; U = \frac{u}{u_m}, \; s = \frac{r_1}{r_2}, \; \Theta = \frac{T - T_o}{T_1 - T_o}.$$

(5)

With the use of Eq. (5), Eq. (1) is satisfied, whereas Eqs. (2–4) take the form as:

$$\left(\frac{\mu_{HNF}}{\mu_F} \right) \left[\frac{d^2 U}{dR^2} + \frac{(1 - s)}{R(1 - s) + s} \frac{dU}{dR} \right] + \left(\frac{(\rho \beta_o)_{HNF}}{(\rho \beta_o)_F} \right) \frac{\beta}{R_e} \cos(\gamma) \Theta - \left(\frac{\sigma_{HNF}}{\sigma_F} \right) H_a^2 \Gamma U$$
$$+ \tilde{\sigma} - \left(\frac{\rho_{HNF}}{\mu_{HNF}} \right) \varepsilon U = 0,$$

(6)

$$\left(\frac{(1 - s)}{R(1 - s) + s} \right) (1 + \kappa\Theta) \frac{d\Theta}{dR} + \kappa \left(\frac{d\Theta}{dR} \right)^2$$
$$+ (1 + \kappa\Theta) \frac{d^2\Theta}{dR^2} + \left(\frac{\mu_{HNF}}{(\rho C_p)_{HNF}} \right) B_r \left(\frac{dU}{dR} \right)^2$$
$$+ \left(\frac{\sigma_{HNF}}{\rho_{HNF}} \right) J(U)^2 + \left(\lambda_{cc} \mathrm{Pr} \frac{k_{HNF}}{(\rho C_p)_{HNF}} \right) \left(U^2 \frac{d^2\Theta}{dR^2} + U \frac{dU}{dR} \frac{d\Theta}{dR} \right) = 0,$$

(7)

with transformed boundary constraints:

$$\text{At } R = 0 \qquad U(R) = \eta_v K_n \frac{dU}{dR}, \; \Theta(R) = 1 + \eta_v K_n \gamma \frac{d\Theta}{dR}.$$

(8)

At $R = 1$

$$U(R) = -\eta_v K_n \frac{dU}{dR}, \quad \Theta(R) = -\eta_v K_n \gamma \frac{d\Theta}{dR}. \tag{9}$$

The dimensionless form of parameters are as follows:

$$H_a = \delta\beta_0 \sqrt{\frac{\sigma_F}{\nu\rho_F}}, \quad \mathrm{Pr} = \frac{(\mu C_p)_F}{k_F}, \quad \tilde{\sigma} = \frac{-\delta^2}{\mu_F} \frac{1}{u_m} \frac{dp}{dr}, \quad \varepsilon = \frac{\delta^2}{\bar{k}},$$

$$\beta = \frac{g(T_1 - T_0)\rho\beta_1\delta^3}{\nu_F^2}, \quad R_e = \frac{u_m\delta}{\nu_F},$$

$$\Gamma = \frac{\cos^2\varpi\,(\beta_e\beta_i + 1)}{(\beta_e\beta_i + 1)(1 + \beta_e\beta_i\cos^2\varpi) + \beta_e^2\cos^2\varpi},$$

$$B_r = \frac{(\delta u_m)^2 \mu_F}{(T_1 - T_0)k_F}, \quad \lambda_{cc} = \frac{\gamma_c u_m^2}{\nu_F}, \quad s = \frac{r_2}{r_1}, \quad J = \frac{u_m^2\beta_0^2\delta^2\sigma_F}{(T_1 - T_0)k_F},$$

$$\eta_v = \frac{2 - \sigma_v}{\sigma_v}, \quad K_n = \frac{\sqrt{\pi R T_0/2\mu_F}}{\rho_F}, \quad \gamma = \frac{B_t}{B_v}. \tag{10}$$

Additionally, Table 2 presents the thermal and physical models for hybrid nanofluids as follows:

2.1 Physical quantities

The equations representing drag force and heat flux at the surface of both inner and outer cylinders are defined as:

$$C_f = \frac{\rho_F \delta^2 \tau_w}{\mu_F^2}, \quad \text{where} \quad \tau_w = \mu_F \frac{du}{dr}, \tag{11}$$

$$Nu = \frac{\delta q_s}{k_F(T_o - T_1)}, \quad \text{where} \quad q_s = -k_{HNF}(T)\frac{dT}{dr}. \tag{12}$$

By utilizing Eq. (5), Eqs. (11) & (12) are transformed into:

For the inner cylinder at $R = 0$, $C_{f_{IN}} R_e = \left(\frac{\mu_{HNF}}{\mu_F}\right)\frac{dU}{dR},$ \tag{13}

$$Nu_{IN} = -\left(\frac{k_{HNF}}{k_F}\right)(1 + \kappa\Theta)\frac{d\Theta}{dR}. \tag{14}$$

Heat transfer analysis

Table 2 Thermal and physical characteristics for hybrid nanofluid flow:.

Properties	Nanofluid
Density (ρ)	$\frac{\rho_{HNF}}{\rho_F} = (1 - \Phi_{s_1} - \Phi_{s_2}) + \Phi_{s_1}\frac{\rho_{s_1}}{\rho_F} + \Phi_{s_2}\frac{\rho_{s_2}}{\rho_F},$
Viscosity (μ)	$\frac{\mu_{HNF}}{\mu_F} = (1 - \Phi_{s_1} - \Phi_{s_2})^{-2.5},$
Thermal conductivity (k)	$\frac{k_{HNF}(T)}{k_F} = \left(\frac{k_{s2} + 2k_{NF} - 2\Phi_{s2}(k_{NF} - k_{s2})}{k_{s2} + 2k_{NF} + 2\Phi_{s2}(k_{NF} - k_{s2})} \times \frac{k_{s1} + 2k_F - 2\Phi_{s1}(k_F - k_{s1})}{k_{s1} + 2k_F + 2\Phi_{s1}(k_F - k_{s1})}\right)(1 + k_1\Theta),$
Heat capacity (ρc_p)	$\frac{(\rho C_p)_{HNF}}{(\rho C_p)_F} = (1 - \Phi_{s_1} - \Phi_{s_1}) + \Phi_{s_1}\frac{(\rho C_p)_{s_1}}{(\rho C_p)_F} + \Phi_{s_2}\frac{(\rho C_p)_{s_2}}{(\rho C_p)_F},$
Thermal expansion $(\rho\beta_1)$	$\frac{(\rho\beta_1)_{HNF}}{(\rho\beta_1)_F} = (1 - \Phi_{s_1} - \Phi_{s_1}) + \Phi_{s_1}\frac{(\rho\beta_1)_{s_1}}{(\rho\beta_1)_F} + \Phi_{s_2}\frac{(\rho\beta_1)_{s_2}}{(\rho\beta_1)_F},$

For the outer cylinder at $R = 1$,

$$C_{f_{OUT}} R_e = \left(\frac{\mu_{HNF}}{\mu_F}\right)\frac{dU}{dR}, \tag{15}$$

$$Nu_{OUT} = -\left(\frac{k_{HNF}}{k_F}\right)(1 + \kappa\Theta)\frac{d\Theta}{dR}. \tag{16}$$

2.2 Numerical analysis

Many nanofluid flow models are addressed by considering various analytical and numerical techniques [20–24]. To compute the results in the form of graphs and tables we have used the bvp4c scheme. This scheme is used to handle highly nonlinear problems. The model in question is approximated to have a tolerance of 10^{-6}. The conversion of Eqs. (6–9) into differential equations of order one is mandatory for numerical computations.

2.3 Outcomes with discussion

This section is devoted to addressing the consequences of dimensionless parameters over the velocity and temperature distributions $(U(R), \Theta(R))$. Fig. 2 is plotted to visualize the role of particle volume fraction of hybrid nanofluid on a velocity distribution $U(R)$. It is seen that the velocity profile enhances against large estimations of (Φ_{HNF}). Physically, increasing the particle volume fraction leads to more nanoparticle collisions, which raises

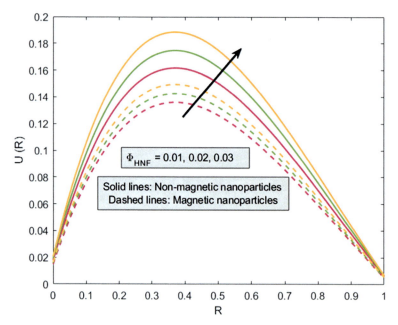

Fig. 2 Influence of nanoparticles volume fraction Φ_{HNF} on the velocity profile $U(R)$.

the velocity profile. It is also worth noting that the velocity of hybrid nanoliquid is higher for nonmagnetic nanoparticles compared to magnetic nanoparticles.

Fig. 3 depicts the impact of the Reynolds number (R_e) on the velocity distribution $U(R)$. For greater counts of Reynolds number, the velocity of hybrid nanoliquid dwindled more for magnetic nanoparticles than nonmagnetic nanoparticles. This is because, when an external magnetic field is present, magnetic nanoparticles experience forces that can impede magnetic nanoparticles-based hybrid nanofluid flow.

Fig. 4 is illustrated to address the effect of the thermal convection parameter (β) on the velocity distribution $U(R)$ of hybrid nanoliquid. This outcome reveals that for large estimations of (β) the velocity distribution increases. Furthermore, this enhancement is shown higher for hybrid nanofluid containing non-magnetic nanoparticles. Heat is transferred employing fluid movement; a process known as thermal convection. Changes in the fluid's velocity distribution may arise from this movement.

Fig. 5 represents the consequences of the Hartmann number (H_a) over the velocity distribution $U(R)$. It is inferred that for greater counts of (H_a), the velocity of hybrid nanofluid containing magnetic nanoparticles

Heat transfer analysis

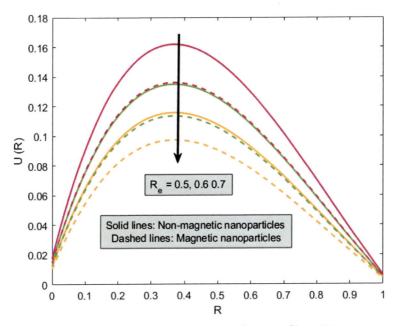

Fig. 3 Influence of Reynolds number R_e on the velocity profile $U(R)$.

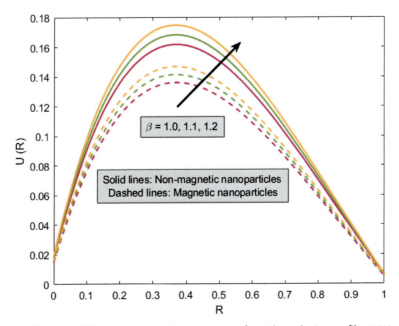

Fig. 4 Influence of thermal convection parameter β on the velocity profile $U(R)$.

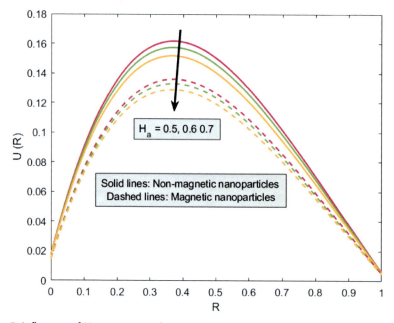

Fig. 5 Influence of Hartmann number H_a on the velocity profile $U(R)$.

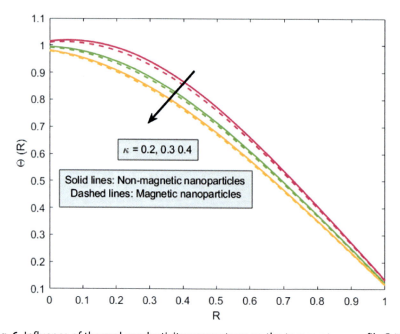

Fig. 6 Influence of thermal conductivity parameter κ on the temperature profile $\Theta(R)$.

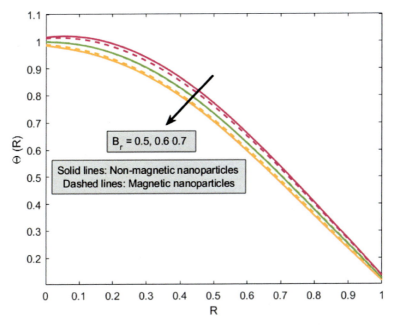

Fig. 7 Influence of Brinkmann number B_r on the temperature profile $\Theta(R)$.

dwindled more significantly. Physically, a resistive force that opposes hybrid nanofluid movement is called the Lorentz force, which arises from the interaction between the magnetic field and the generated electric currents in the fluid.

Fig. 6 is portrayed to show the role of variable thermal conductivity parameter (κ) over the temperature distribution $\Theta(R)$. It is revealed that the temperature distribution of hybrid nanoliquid containing magnetic nanoparticles decreases more against (κ). This is because, for mounting values of variable thermal conductivity parameter (κ), the heat transmission rate increases which consequently diminishes the temperature distribution of hybrid nanofluid flow.

Moreover, Fig. 7 is sketched to explore the trend of temperature distribution $\Theta(R)$ against the Brinkmann number (B_r). It is seen that temperature distribution dwindled against the greater counts of (B_r). Physically, the Brinkmann number upsurges the heat transmission rate which lowers the temperature distribution.

Lastly, to visualize the consequences of the Joule parameter (J) over the temperature distribution $\Theta(R)$, Fig. 8 is displayed. It appears that the temperature distribution improves more frequently for hybrid nanofluid

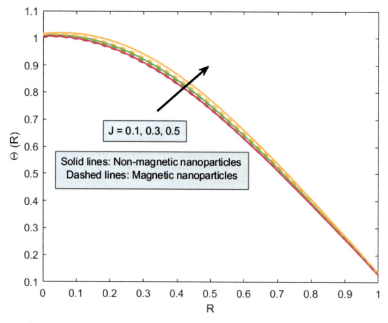

Fig. 8 Influence of Joule parameter J on the temperature profile $\Theta(R)$.

flow containing magnetic nanoparticles. From a physical point of view, a larger Joule parameter means that Joule heating is more pronounced in the fluid, and this might raise the temperature profile.

Furthermore, Table 3 is drawn to visualize the trend of the surface drag coefficient $(C_{f_{IN}})$ at the inner cylinder against particle volume fraction (Φ_{HNF}), Joule parameter (J) and Reynolds number (R_e). It is revealed that for greater counts of (Φ_{HNF}), the surface drag is enhanced, and this augmentation is dominating for hybrid nanofluid containing magnetic nanoparticles. Whereas raising the Joule parameter (J) and Reynolds number (R_e) raises the surface drag coefficient, this increase is more profound for hybrid nonliquids containing non-magnetic nanoparticles. Table 4 illustrates the role of distinct factors including nanoparticle volume fraction (Φ_{HNF}), Joule parameter (J) and Reynolds number (R_e) over heat transmission rate (Nu_{IN}). It is revealed that heat transmission is greater for hybrid nanofluid containing magnetic nanoparticles when the Joule parameter (J) increases whereas for mounting estimates of (Φ_{HNF}) and (R_e) the heat transmission rate (Nu_{IN}) at the inner cylinder dwindled more for hybrid nanofluid containing non-magnetic nanoparticles.

Table 3 Numerical estimations for surface drag coefficient $(C_{f_{IN}})$ at inner cylinder for varied parameters for magnetic/non-magnetic hybrid nanofluids.

(Φ_{HNF})	(J)	(R_e)	Magnetic nanoparticles	Nonmagnetic nanoparticles
			$(C_{f_{IN}})$	
0.020	1	0.5	0.952514	1.14538
0.025	–	–	0.962189	1.23393
0.030	–	–	0.976612	1.32729
–	1.1	–	0.935105	1.11503
–	1.2	–	0.924370	1.09696
–	–	0.6	0.794105	0.954771
–	–	0.7	0.680840	0.818532

Table 4 Numerical estimations for surface heat transmission rate (Nu_{IN}) at inner cylinder for varied parameters against magnetic/non-magnetic hybrid nanofluids.

(Φ_{HNF})	(J)	(R_e)	Magnetic nanoparticles	Non-magnetic nanoparticles
			(Nu_{IN})	
0.020	1	0.5	0.405393	0.388865
0.025	–	–	0.397914	0.376807
0.030	–	–	0.390031	0.363835
–	1.1	–	0.406345	0.390270
–	1.2	–	0.407296	0.391670
–	–	0.6	0.403408	0.386200
–	–	0.7	0.402209	0.384593

2.4 Future areas of research

- Temperature-dependent viscosity and electrical and thermal conductivities can be incorporated.
- A comparison of coolants-based hybrid nanofluid flow can be done.
- Cattaneo–Christov heat flux can be incorporated.

3. Conclusions

This study has focused on the consequences of Hall and ion slip parameters on the magnetic/non-magnetic hybrid nanofluid flow in an annular region of two slanted inclined concentric cylinders in a porous medium. Two different hybrid nanofluids are considered, one with Silver (Ag) and Copper (Cu), as non-magnetic nanoparticles whereas cobalt ferrite $(CoFe_2O_4)$ and $(Mn - ZnFe_2O_4)$ are considered as magnetic nanoparticles immersed in water (H_2O) which is assumed as a baseliquid. Heat transmission is analyzed while considering viscous-ohmic dissipations, temperature-dependent thermal conductivity, and Cattaneo-Christov heat flux. The salient results are outlined as under:

- With greater counts of particle volume fraction, the velocity distribution is further enhanced in hybrid nanofluids containing non-magnetic nanoparticles.
- Velocity distribution for hybrid nanofluid containing magnetic nanoparticles declined for mounting counts of Hartmann number.
- Temperature distribution dwindled for increasing estimations of variable thermal conductivity parameter whereas it enhances for increasing Joule parameter.
- Surface drag coefficient is experienced more for hybrid nanofluid containing non-magnetic nanoparticles.
- Heat transmission rate upsurges more for hybrid nanofluid containing magnetic nanoparticles.

References

[1] S.U. Choi, J.A. Eastman, Enhancing thermal conductivity of fluids with nanoparticles (No. ANL/MSD/CP-84938; CONF-951135-29). Argonne National Lab (ANL), Argonne, IL, 1995.

[2] J.A. Eastman, S.U.S. Choi, S. Li, W. Yu, L.J. Thompson, Anomalously increased effective thermal conductivities of ethylene glycol-based nanofluids containing copper nanoparticles, Appl. Phys. Lett. 78 (6) (2001) 718–720.

[3] N. Shaheen, M. Ramzan, C.A. Saleel, S. Kadry, Analysis of Newtonian heating and surface catalyzed reaction in a trihybrid nanofluid flow across an expanding/shrinking cylinder with Thompson and Troian slip, Proc. Inst. Mech. Eng. N: J. Nanomater, Nanoeng Nanosyst. (2024) 23977914231225174.

[4] N. Shahmir, M. Ramzan, C.A. Saleel, S. Kadry, A comparative assessment of mono and hybrid magneto nanofluid flow over a stretching cylinder in a permeable medium with generalized Fourier's law, Numer. Heat. Transf. A: Appl. (2023) 1–19.

[5] H. Gul, M. Ramzan, C.A. Saleel, S. Kadry, A comparative analysis of ternary-hybrid nanofluid flows through a stretching cylinder influenced by an induced magnetic field with homogeneous–heterogeneous reactions, Numer. Heat. Transf. A: Appl. (2023) 1–16.

[6] A. Alsaedi, K. Muhammad, T. Hayat, Numerical study of MHD hybrid nanofluid flow between two coaxial cylinders, Alex. Eng. J. 61 (11) (2022) 8355–8362.

[7] T. Tayebi, A.J. Chamkha, Free convection enhancement in an annulus between horizontal confocal elliptical cylinders using hybrid nanofluids, Numer. Heat. Transf A: Appl. 70 (10) (2016) 1141–1156.

[8] N. Shahmir, M. Ramzan, C.A. Saleel, S. Kadry, Impact of induced magnetic field on ferromagnetic ternary and hybrid nanofluid flows with surface catalyzed reaction and entropy generation assessment, Numer. Heat. Transf A: Appl. (2023) 1–21.

[9] K.A.M. Alharbi, J. Ali, M. Ramzan, S. Kadry, A.M. Saeed, A comparative analysis of hybrid nanofluid flow through an electrically conducting vertical microchannel using Yamada-Ota and Xue models, Numer. Heat. Transf A: Appl. (2023) 1–16.

[10] Z. Mahmood, Z. Duraihem, F. Adnan, U. Khan, A.M. Hassan, Model-based comparative analysis of MHD stagnation point flow of hybrid nanofluid over a stretching sheet with suction and viscous dissipation, Numer. Heat. Transf, B: Fund (2024) 1–22.

[11] S. Dinarvand, M. Nademi Rostami, M. Yousefi, S.M. Mousavi, B. Jabbaripour, S. Noeiaghdam, Reiner-Philippoff model accompanied by a mass-based algorithm for hybrid bio-nanofluid flow over a nonlinearly-mutable wall, Numer. Heat. Transf, B: Fund (2024) 1–14.

[12] R. Razzaq, U. Farooq, Non-similar analysis of MHD hybrid nanofluid flow over an exponentially stretching/shrinking sheet with the influences of thermal radiation and viscous dissipation, Numer. Heat. Transf, B: Fund (2024) 1–16.

[13] A. Rajeev, V. Puneeth, S. Manjunatha, O.D. Makinde, Magnetohydrodynamic flow of two immiscible hybrid nanofluids between two rotating disks, Numer. Heat. Transf, A: Appl. (2024) 1–19.

[14] K. Goyal, S. Srinivas, Pulsatile flow of Casson hybrid nanofluid between ternary-hybrid nanofluid and nanofluid in an inclined channel with temperature-dependent viscosity, Numer. Heat. Transf, A: Appl. (2024) 1–30.

[15] M. Iqbal, N.S. Khan, W. Khan, S.B.H. Hassine, S.A. Alhabeeb, H.A.E.W. Khalifa, Partially ionized bioconvection Eyring–Powell nanofluid flow with gyrotactic microorganisms in thermal system, Therm. Sci. Eng. Prog. 47 (2024) 102283.

[16] A. Salmi, H.A. Madkhali, M. Nawaz, S.O. Alharbi, A.S. Alqahtani, Numerical study on non-Fourier heat and mass transfer in partially ionized MHD Williamson hybrid nanofluid, Int. Commun. Heat. Mass. Transf. 133 (2022) 105967.

[17] M. Ramzan, N. Shahmir, H.A.S. Ghazwani, Mixed convective Casson partially ionized nanofluid flow amidst two inclined concentric cylinders with gyrotactic microorganisms, Waves Random Complex. Media (2022) 1–21.

[18] P.M. Patil, H.F. Shankar, Heat transfer attributes of Al_2O_3-Fe_3O_4/H_2O hybrid nanofluid flow over a yawed cylinder, Propuls. Power Res. 11 (3) (2022) 416–429.

[19] J.D. Chung, M. Ramzan, H. Gul, N. Gul, S. Kadry, Y.M. Chu, Partially ionized hybrid nanofluid flow with thermal stratification, J. Mater. Res. Technol. 11 (2021) 1457–1468.

[20] S. Li, V. Puneeth, F.A. Al-Yarimi, S. Manjunatha, M.S. Anwar, A.J. Chamkha, Thermal and solutal stratified Heimanz flow of AA 7072-deionized water over a wedge in the presence of bioconvection, Numer. Heat. Transf. B: Fund (2024) 1–19.

[21] A. Hussain, Z. Mao, Heat transfer analysis of MHD Prandtl-Eyring fluid flow with Christov-Cattaneo heat flux model, Numer. Heat. Transf, Part. A: Appl. (2024) 1–21.

[22] K. Leelasagar, K. Ventaktasubbaiah, Numerical investigation of natural convection flow inside a square enclosure filled with different nanofluids by using two-phase Eulerian–Eulerian model: a new correlation for Nusselt number, Numer Heat Transf, A: Appl. in press, 2024, https://doi.org/10.1080/10407782.2024.2316209.

[23] J. Madhu, K. Karthik, R.S. Varun Kumar, R.J. Punith Gowda, R. Naveen Kumar, B.C. Prasannakumara, Dynamics of pollutant dispersion and solid–fluid interfacial layer in Jeffrey nanofluid flow subjected to waste discharge concentration: implementation of probabilists' Hermite polynomial collocation method, Numer. Heat. Transf A: Appl. (2024), https://doi.org/10.1080/10407782.2024.2319349

[24] I. El Glili, M. Driouich, Pulsatile flow of EMHD non-Newtonian nano-blood through an inclined tapered porous artery with combination of stenosis and aneurysm under body acceleration and slip effects, Numer. Heat. Transf A: Appl. (2024) 1–29.

Printed in the United States
by Baker & Taylor Publisher Services